T0329600

CD files for this title can now be found by entering the ISBN

9780471655312

on booksupport.wiley.com

LC/MS

LC/MS

A Practical User's Guide

MARVIN C. McMASTER

WILEY-
INTERSCIENCE

A JOHN WILEY & SONS, INC., PUBLICATION

Published by John Wiley & Sons, Inc., Hoboken, New Jersey.
Published simultaneously in Canada.

Limit of Liability/Disclaimer of Warranty: While the publisher and author have used their best efforts in preparing this book, they make no representations or warranties with respect to the accuracy or completeness of the contents of this book and specifically disclaim any implied warranties of merchantability or fitness for a particular purpose. No warranty may be created or extended by sales representatives or written sales materials. The advice and strategies contained herein may not be suitable for your situation. You should consult with a professional where appropriate. Neither the publisher nor author shall be liable for any loss of profit or any other commercial damages, including but not limited to special, incidental, consequential, or other damages.

For general information on our other products and services or for technical support, please contact our Customer Care Department within the United States at (800) 762-2974, outside the United States at (317) 572-3993 or fax (317) 572-4002.

Wiley also publishes its books in a variety of electronic formats. Some content that appears in print may not be available in electronic formats. For more information about Wiley products, visit our web site at www.wiley.com.

Library of Congress Cataloging-in-Publication Data:

McMaster, Marvin C.
    LC/MS: a practical user's guide / Marvin C. McMaster.
       p. cm.
    Includes bibliographical references and index.
    ISBN-13 978-0-471-65531-2 (cloth)
    ISBN-10 0-471-65531-7 (cloth)
    1. Liquid chromatography—Handbooks, manuals, etc. 2. High performance liquid chromatography—Handbooks, manuals, etc. 3. Mass spectrometry—Handbooks, manuals, etc. I. Title.
    QD79.C454M363 2005
    543'.84—dc22

                                                    2004063820

10 9

To the memory of my son, Chris McMaster, my writing partner and the artist on the first two books in this series. Chris has passed on to bigger and better things painting sunrises and rainbows.

CONTENTS

PREFACE

I consult and teach extension courses on laboratory instrumentation and computers at the University of Missouri–St. Louis. I taught a course called *Practical HPLC* for a number of years while working as a sales representative and technical support specialist for a variety of instrument companies. The first book in this series, *HPLC: A Practical User's Guide*, arose out of a need for a textbook for my course. At the end of that book I wrote a chapter on a rising research technique that I felt would eventually transform the life of the average laboratory chemist and provide a tool for definitive identification of the compounds that he or she was producing.

I next had an opportunity to work with a manufacturer of control and data systems for GC/MS equipment. I added consulting and teaching in this specialty to my portfolio and designed a book, *GC/MS: A Practical User's Guide*, to provide a teaching tool. Again, I added a final chapter on the growing art of LC/MS. I feel another book and course are needed now that commercial sales of LC/MS systems has nearly equaled those of GC/MS systems. This tool combines my expertise and interests in several separations areas.

I do not attempt to write the definitive book for a new instrumentation specialty. I want to put together a useful tool for introducing the technique and providing practical information on how to use it. I try to look at complicated material, internalize it, and present it in a way that is understandable and useful for solving laboratory problems. When inexpensive, easy-to-use LC/MS systems appear on the end of every laboratory bench, I would like to have a copy of this book setting next to them to lay the groundwork for getting the most out of the system.

When I teach practical courses, I use an overhead projector and a PowerPoint slide set to provide the theme and illustrations for the course. I realize that if I were buying this book to use as a teaching text book, it would be very useful to have the slide set on a CD/ROM disk. In the back of this book I have included such a disk with my slide set, searchable files on LC/MS Frequently Asked Questions, a glossary of terms, and useful LC/MS tables. For the LC/MS students, this provides a series of self-study guides for learning or honing their LC/MS skills. I hope the readers of this book will find these additional tools useful. I plan to add similar tools to later editions of my other books.

I wish to thank the following companies for permission to use drawings and illustrations from their brochures and Web sites: Agilent Technologies, Applied Biosystems, ESA, Varian, and Waters Corporation. I have found in teaching that pictures truly are worth a thousand words. Their kind assistance has helped me keep this book down to a reasonable size. I never have cared for "rat killer" manuals.

MARVIN C. MCMASTER

Florissant, Missouri

1

INTRODUCTION TO LC/MS

Liquid chromatography (LC) combined with mass spectrometry (MS) creates an ideal analytical tool for the laboratory. The high-performance liquid chromatograph (HPLC) has been the laboratory tool of choice for separating, analyzing, and purifying mixtures of organic compounds since the 1970s.

An HPLC column can separate almost any mixture that can be dissolved. A mass spectrometer can ionize the separated peak solution and provide a molecular weight for each peak component. An LC/MS/MS system can fragment the parent ion into a distinctive fragmentation pattern and can separate the daughter ions for identification and quantitation. The characteristic fragmentation pattern from each parent ion can be identified by comparison to fragmentation patterns produced by standard computerized databases. The output of the HPLC system can be divided for analysis by other HPLC detectors or for preparative sample recovery, since only a small portion of the column effluent is required for mass spectral analysis.

1.1 WHY LC/MS?

The preferred tool until the turn of the millennium for separating a mixture and providing definitive identification of its components was the gas chromatograph/mass spectrometer (GC/MS). However, this technique was limited by three main factors:

1. Sample volatility.
2. The fact that aqueous samples require extraction.
3. Thermal degradation of samples in the GC oven.

LC/MS: A Practical User's Guide, by Marvin C. McMaster
Copyright © 2005 John Wiley & Sons, Inc.

Not all compounds are volatile enough to be introduced or eluted off a GC column. Aqueous mixtures have to be extracted and/or derivatized before injection, adding to analysis cost and bringing sample handling errors into peak quantitation. The columns available were not able to resolve all mixtures of compounds. This problem has been eliminated somewhat with new varieties of columns. Oven-temperature programming remains the principal variable available for separating compounds in a mixture. The final oven temperature necessary to remove a large compound from a column can degrade many thermally labile compounds.

In the last two years, LC/MS sales have nearly equaled GC/MS sales because of the additional compounds that can be analyzed by LC/MS and the greater range of separation variables that can be utilized in HPLC separation. The editors of *Analytical Instrumentation Industry Reports* say that in 2000 the global GC/MS market was $300 million and that LC/MS sales reached $250 million. This does not indicate parity, but it does show that the gap is closing. One industry analyst predicted that LC/MS sales should top $1 billion by 2005. The difference in cost of a HPLC system and its interface compared to a gas chromatograph must be factored into these numbers when comparing unit costs. An isocratic HPLC system costs 50% more than a basic GC module. The cost difference nearly doubles when you add in the cost of an atmospheric-pressure interface (API). Gradient HPLC configuration increases the cost to triple that of a GC module. However, all of these costs are overshadowed by the price of a mass spectrometer.

For LC/MS to be a major player in the analytical laboratory, there are factors limiting performance that must be overcome:

- Analyzer signal swamping by the elution solvent.
- Solvent composition changing in gradient elution.
- Buffer use for pH control.
- Ionization of neutral peak components.

By far the most important of these is the volume of eluting solvent necessary to displace the compounds separated from the HPLC column. The mass analyzer is quickly overwhelmed by the signal from the solvent if the HPLC output is introduced directly into the mass spectrometer. The analyate signal is buried beneath this solvent signal avalanche. The solvent signal saturation effect occurs even if a low-molecular-weight solvent such as methanol or water is chosen and a low analyzer mass cutoff range is selected to exclude the solvent's peak signal. A method for in-stream solvent removal with concurrent sample concentration must be provided to connect the column effluent to a high-vacuum mass spectrometer. The HPLC solvent gradient used to resolve closely eluting HPLC peaks and decrease HPLC run times also produces solvent composition changes that further complicate the solvent-masking effect of analyate signal.

Many compounds resolved by the HPLC column require pH control to adhere to the column long enough to be eluted. Removal of nonvolatile buffer and ion-pairing reagents commonly used in HPLC separations from the effluent is the next problem that must be handled. Direct introduction of inorganic compounds into a high-vacuum system will cause mass spectrometer inlet fouling and loss of signal. Organic buffers used instead of inorganic buffers exhibit the same problems as those found with organic solvents: They overwhelm the analyzer and detector. Replacing nonvolatile buffers and reagents with volatile equivalents allows them to be removed like solvent. The final hurdle is that neutral compounds separating off the HPLC column must be converted to charged molecular ions or fragmented into charged ions that can be separated by the analyzer.

API using ion spray and electrospray interfaces provides many of the answers to these problems. At least part of the stream from the HPLC is sprayed over a high-voltage coronal discharge needle in a heated chamber, vaporizing the solvent and charging the suspended molecule, creating a molecular ion. A neutral flowing curtain gas sweeps much of the solvent and volatile additives out of the interface before the ionized analyate is pulled into the pinhole entrance to the high-vacuum environment of the analyzer. One of Jack Henion's papers produced at Cornell University reports that he operated an ion spray interface at effluent flow rates of 2 mL/min of methanol/water containing phosphate buffer to feed sample into a Hewlett-Packard MSD mass spectrometer with its vacuum provided by a tiny turbo pump, but this should be looked on as an exception to the rule of using volatile components.

Liquid chromatography provides a wide variety of operating variables that can be used to control and optimize a separation:

- Column-bonded phase selection with rapid column switching.
- Major solvent change with rapid reequilibration.
- Mobile-phase polarity adjustment and gradient operation.
- Packing support selection for pH and temperature stability.
- Temperature programming.

Most HPLC separations have been carried out using reverse-phase silica columns, with non-polar-bonded phases eluting compounds with polar solvents. A wide variety of bonded phases are available to achieve these separations. Various nonpolar mobile-phase solvents can be selected to shift elution orders of compounds on the same type of column. Mixing nonpolar solvents with water can change solvent polarity, increasing or decreasing partitioning with bonded-phase packing.

Traditional HPLC column supports have had nonpolar bonded phase bound to a silica matrix. These bonded phases are unstable under strongly acidic conditions, and the silica matrix dissolves rapidly at mildly basic pH. Newer polymeric and zirconium matrixes provide reverse-phase columns that are both pH and temperature stable. These packing materials allow operation at high or low pH without using buffers. Zirconium packing allows use of temperature as a separations variable using a temperature-controlled column jacket. Thermally labile compounds would have some of the same problems as those seen in a GC oven, but the temperature control range is much lower in HPLC, due to solvent volatility.

In the first section of this book we focus on optimization of the liquid chromatograph. We discuss equipment configurations, columns, and separation variables that can help improve peak resolution. Routine maintenance tips will show how to maintain the system and the separation without contaminating the interface and the mass spectrometer. An earlier book in this series, *HPLC: A Practical User's Guide*, provides additional information on using and optimizing the performance of silica-based HPLC columns.

In the second section of the book we look at the components that make up the various analyzers used in mass spectrometry. We compare the advantages and areas of specific applications of quadrupole, ion trap, Fourier transform, and time-of-flight (TOF) analyzer configurations. A variety of systems for generating the high vacuum used in analyzer operations are described and evaluated. Techniques for maintain a system under operating conditions and for cleaning contaminated analyzers are explained. The basic theory for controlling analyzer and detector sensitivity and scanning ranges is discussed. Two of the great advances in interpretation of mass spectral data have been the introduction of accurate mass-molecular-weight determination and computer scanning of library databases of known fragmentation patterns to aid compound identification. These have greatly reduced the time and operator skills needed to use and understand information generated by mass spectrometers. A brief introduction to fragmentation pattern interpretation from LC/MS/MS data is provided to aid in checking database search results.

1.2   MOLECULAR WEIGHTS AND STRUCTURE STUDIES

In the final section of the book we look at a number of current application areas for LC/MS in biochemical and industrial laboratories as well as other areas of LC/MS application that are anticipated when regulated methods become available. Special emphasis is placed on drug discovery and development, protein analysis, impurities, and metabolite determinations. These areas have fueled the rapid growth of LC/MS sales in the last few years. The needs of these labs go beyond the desire to provide separation and molecular weights for compounds in synthesis mixtures. Fragmentation studies using LC/MS, LC/MS/MS, and mixed analyzer systems to supplement LC/MS comprise a rapidly growing technology. It is important to understand the changes in system costs, hardware configurations, applications, and techniques that seem to be driving these changes.

1.3   LC/MS SYSTEMS

Basically, an LC/MS system is an HPLC pumping system, injector, and column married to a mass spectrometer through some type of evaporative ionizing interface (Figure 1.1). A computer system coordinates the components of the system together by providing control of the HPLC for flow, solvent gradient, and remote starting of injection and the gradient run. It also provides control of the mass

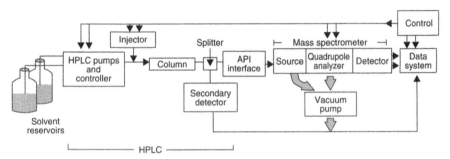

FIGURE 1.1 LC/MS system model.

spectrometer scan range and lens, and accesses and processes data from the ion detector's amplifier. All of this is done through either a remote control interface or through A/D (analog-to-digital; data input) and D/A (digital-to-analog; control) microprocessor cards in the computer system module. The digital data from the A/D card is then processed by the computer's software to provide a total ion chromatogram (TIC) and the molecular weights of the compounds in the peaks detected using the mass spectrometer's spectral data. Obviously, the computer is very busy and requires the very latest in processor speed, memory, and data storage.

The system I have described sounds complex but is in reality a very basic LC/MS system and provides only basic information. We will need a complex LC/MS/MS system if we want more information to identify the compound of interest (Figure 1.2). Such a system measures not only molecular weights but can also fragment the precursor ion provided by the first separation into smaller ions and measure the molecular weights of these ions by doing a product mass scan. We can use this information to develop a structural interpretation of the original structure either by rigorous deduction from fragmentation peak positions

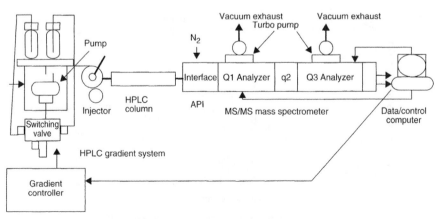

FIGURE 1.2 LC/MS/MS triple-system model.

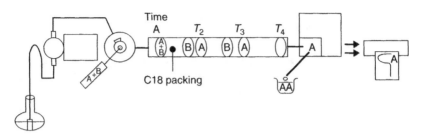

FIGURE 1.3 Column separation model.

or by computer comparison to a commercial database of known compounds and their fragmentation patterns—one more job for the overworked system computer.

All of this hardware and software is in place to run a metal column packed with highly particulate material. Solvent from the pumping system is forced through the HPLC column, and dissolved sample is injected into the flowing stream. The material dissolved in the mobile phase interacts with the packing material, and equilibration separation occurs as the material moves down the column (Figure 1.3).

Disks of these separated compounds elute off the column at different times, enter the interface where solvent is evaporated and the compounds are ionized, and are then pulled into the evacuated mass spectrometer. Electrical lenses focus the charged beam of ions and carry them into the mass analyzer. They are swept down the analyzer by a scanning direct-current/radio-frequency (dc/RF) signal that selects ions of a particular mass/charge (m/z) value to strike the

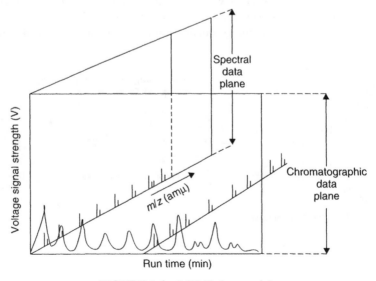

FIGURE 1.4 LC/MS data model.

detector face and trigger a signal. This signal is combined in the computer with control information that it is sending to the mass spectrometer to create a three-dimensional array of signal strength versus time versus m/z information for storage and processing (Figure 1.4).

1.4 SYSTEM COSTS

System prices are very difficult to gather from equipment manufacturers; they guard them like a mother hen protecting her chicks. What I have put together are simply estimates obtained by talking to customers. A basic four-solvent gradient quadrupole ESI (electrospray interface)-LC/MS with its control computer intended for molecular-weight determination would cost approximately $140,000. I talked recently with an employee at an HPLC company that had just purchased a Qtrap LC/MS/MS system for $220,000. A university group setting up a core facility told me they had a bid of $750,000 for a MALDI (maser-assisted laser desorption and ionization)/TOF LC/MS and LC/Qtrap MS/MS system with a protein database system. This included a two-dimensional electrophoresis system to do two-dimensional protein gels and a robotic laboratory setup. I also talked to a university group that had retrofitted a Hewlett PACKARD 5971 MSD mass spectrometer from a GC/MS that had been purchased originally for $86,000 with an $18,000 three-solvent gradient HPLC and a $12,000 ion spray interface. Getting started in LC/MS is not a casual adventure.

1.5 COMPETITIVE SYSTEMS

HPLC is not the only separation system being used as a front end for mass spectral analysis. Applications using GC/MS preceded LC/MS by a number of years and are very common in environmental and toxicology laboratories, where standard methods for their use exist, provided by agencies such as the U.S. Environmental Protection Agency and the Association of Analytical Chemists. A GC/MS requires a sample that is volatile or can be derivatized and is thermally stable under the column conditions used for separation. A model GC/MS system is shown in Figure 1.5.

Capillary zone electrophoresis (CZE) has proved to be a powerful separation and analysis tool. Ionized samples in buffer are forced through a partition gel packed capillary column down a voltage potential applied over the length of the column and are eluted into the mass spectrometer interface. CZE/MS continues to gain popularity but lacks the versatility of HPLC's wide range of column types and control variables (Figure. 1.6).

The final candidate for mass spectrometer upgrades is supercritical fluid (SCF) chromatography. This technique is popular in the flavor, perfume, and essential oil manufacturing sectors. It uses gases such as carbon dioxide, methane, and ammonia as liquids above their supercritical pressure and temperature point as

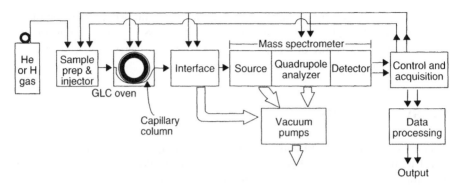

FIGURE 1.5   GC/MS model system.

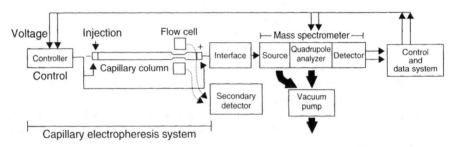

FIGURE 1.6   CZE/MS model system.

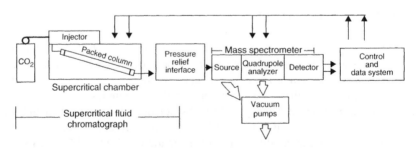

FIGURE 1.7   SCF/MS model system.

the mobile phase on conventional HPLC columns. Interfaced into an ion spray interface and a mass spectrometer, they create an SCF/MS system (Figure 1.7). This is an interesting system for preparation purposes; simply releasing the pressure and letting the working fluid evaporate allows the separated compounds to be recovered.

2

THE HPLC SYSTEM

The LC part of an LC/MS system is made up of the hardware and column of an HPLC system. A basic LC/MS configuration would be made up of a solvent pump, a sample injector, an HPLC column, a detector, a data collection component, and small-diameter tubing to connect all the liquid components (Figure 2.1). Some provision must be made to acquire the signal from the detector to provide a record of the separation achieved in the column. This might be either a stripchart recorder or an integrator, but today it would probably be a data acquisition module within a computer. Finally, if the effluent from the column is to be taken directly to the mass spectrometer, an interface must be provide to remove volatile mobile-phase components and to ionize the peak components.

2.1 HPLC SYSTEM COMPONENTS

The heart of the HPLC system is the column where the actual separation occurs. A mobile phase is pumped from a reservoir, through an injector, into the column, and out to the detector. A sample dissolved in the mobile phase or a similar solvent is injected into the flowing mobile phase on the column, separation occurs that is specific for that type of column, and the separated peak elute flowing into the detector causes a signal to be sent to the data system. We will leave discussion of the various types of columns and separation modes to the next chapter and focus here on the hardware that supports the column.

LC/MS: A Practical User's Guide, by Marvin C. McMaster
Copyright © 2005 John Wiley & Sons, Inc.

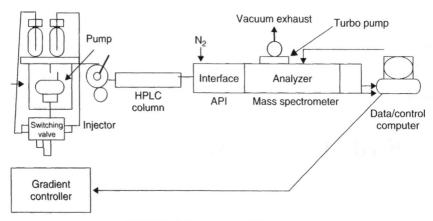

FIGURE 2.1   Basic LC/MS system.

Let's start with the first hardware component, the HPLC pump. The pump takes in solvent from a reservoir through some type of filter, pressurizes the solvent sufficiently to overcome resistance from the column packing, and drives the solvent into the injector. Solvents are drawn into the pump by suction, and it is important that they be degassed before they are placed in the solvent reservoir unless the system is designed to degas solvents automatically. Degassing can be done by sonication, but the most effective degassing method is suction filtration through a fine-pore-size fritted filter. A third method of degassing involves sparging the solvent with an inert gas such as helium. The reciprocating piston displacement pump is the most commonly used HPLC pump. It consists of a metal body drilled out to provide a pumping chamber that is sealed at the back with a Teflon seal through which rides an inert piston. Check-valve-equipped inlet and outlet ports allow solvent to enter and exit the pumping chamber (Figure 2.2).

The inlet check valve closes to prevent solvent from running back into the solvent reservoir during the pressurization portion of the piston stroke. At the same time, the outlet check valve pops open to allow solvent delivery to the line leading to the injector. When the cam-driven motor pulls the piston back, the inlet check valve pops open to admit more solvent while the outlet check valve closes to prevent runback of pressurized solvent from the injector line.

The keys to the operation of the pump are the piston and the piston seal. The piston must be resistant to corrosion by the solvent components, which may include high salt concentrations used in ion-exchange columns and 6 normal nitric acid used to clean and pacify extracolumn wetted surfaces. The most commonly used pump pistons are made of beryl glass and are commonly referred to as *sapphire pistons*. Sapphire pistons are not blue, by the way, but the name helps justify the cost when a broken one has to be replaced. Pistons have great strength along their drive axis but are easily snapped across the axis. Most pumps are designed to avoid piston wobble, so the most common reasons for breaking a piston are buffer buildup on the seal and breakage when pump heads are being removed to check the condition of the piston (Figure 2.3).

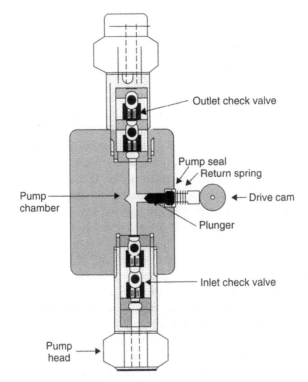

FIGURE 2.2 HPLC pump head.

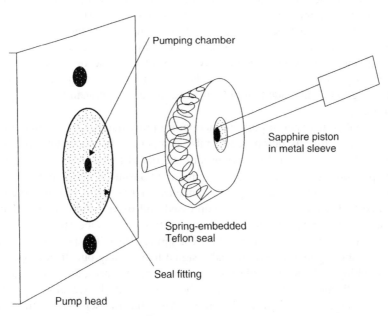

FIGURE 2.3 Piston and seal.

The seal is a marvel of construction critical to pump operation. It is a torus of Teflon containing an embedded circular steel spring with a hole in the center through which the pump piston passes. This doughnut-shaped seal fits in a circular depression at the back of the pump head, not quite as deep as the thickness of the seal. The seal is compressed and the spring squeezes onto the pump piston when the pump head is fastened to the pump face with screws. It creates a high-pressure liquid barrier around the piston as it rides forward and backward in its stroke. Liquid from the pumping chamber lubricates the piston as it rides through the Teflon seal and evaporates as the film on the piston reaches the outside of the seal. Buffers or ion-pairing reagents in the mobile phase crystallize on the piston and are wiped off by the seal on the return stroke. If they do not wipe off or are not washed off, they accumulate and turn the piston into a saw that cuts through the Teflon, producing leaks which require that the seal be replaced periodically. Seal replacement is the operation that most commonly causes piston breakage if not done correctly.

The drive cam and the pressure transducer are two other components that influence pump operation. The basic problem with this type of pump is that it pulses. Part of the time the pump is pressurizing and driving solvent toward the column, and part of the time the piston is refilling the solvent chamber. Pressure in the pumping chamber rises and falls, resulting in pulses of solvent delivered through the outlet check valve. This problem is overcome by three methods: use of opposing multiple pump heads, electronic pump motor control, and pulse damping. Pulsing is reduced if you have two pump heads feeding the same solvent line through a T-tube at different times. One can be refilling while the other is driving out solvent. Obviously, this method increases the cost of the pump by adding components and engineering, but it does produce the best solvent delivery. By controlling the pump motor electronically to speed up in the refill and repressurization stroke, you can design a single-piston pump that spends the majority of its time in the delivery mode. This pump still pulses, but the pulsing is reduced dramatically. It does not perform as well as a good dual-headed pump performs, but it is significantly less expensive to build.

The final component in pulse reduction is a pulse dampener. No manufacturer likes to admit that its pump needs pulse dampeners, but all manufacturers use them. A device with two lines going into and out of a metal can in-line between a pump's outlet valve and injector is a pressure transducer, a pressure sensor, or a pulse dampener. Cut open the pulse dampener on a high-pressure pump and you will find that it contains a long, compressed coil of very fine-internal-diameter stainless steel tubing. When a pulse occurs, this coiled tube stretches and then compresses again, damping the pulse by a spring effect. The pressure transducer also has tubing going in and out and a signal line coming out. Inside is a curved coil of tubing with an attached sensor. As pressure increases, the tubing stretches, and this deflection can be measured by the sensor, with the signal being sent to a pressure gauge on the front of the pump. When making separations that require inert conditions, it is important to understand that these devices are present. Both contain stainless steel that can corrode by high concentrations of salts and can contaminate metal-sensitive enzyme purifications.

The second hardware component in the HPLC system is the injector. The loop-and-valve injector is the most commonly used type of injector. In one position of the valve, the high-pressure liquid flow from the pump flows through an internal bypass and out to the HPLC column. The other side of the valve is connected to a small-diameter loop, an injection port, and an overflow drain, as shown in Figure 2.4a.

Sample dissolved in the mobile phase is injected into the loop from a syringe, overfilling the loop by at least 20%. It is also possible to use a partial injection as long as the sample is loaded slowly and kept to no more than 75% of the loop volume to avoid losing sample out of the overflow port. Next, the injector

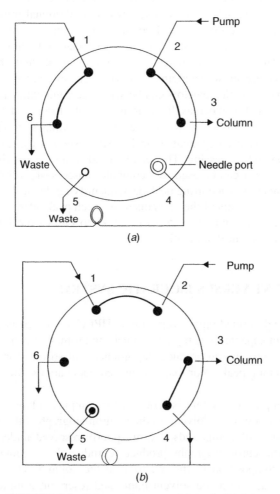

FIGURE 2.4 Injector diagrams: (a) load; (b) inject.

handle is turned rapidly to the inject position (Figure 2.4*b*). The sample is washed out from the back end of the loop, given a last in/first out injection to maintain sample concentration. Autosamplers generally employ this type of loop and valve arrangement, although they may pull the sample into the loop instead of pushing it in. They also use a variety of wash schemes to avoid sample carryover contamination when moving from vial to vial for subsequent injections.

The third hardware component in the HPLC systems is the detector or detectors. In the LC/MS system this would be one of the mass spectrometer detectors, but other HPLC detectors may be used on the same solvent stream from the column, either in series with the mass spectrometer or by using a splitter to divert part or most of the effluent to a secondary detector such as a refractive index (RI) detector, conductivity detector, ultraviolet (UV) detector, or fluorometer (FL). All of these secondary peak detectors must use very low-dead-volume flow cells if they are to keep from remixing the separated chromatography peaks or diluting them with mobile phase. Each provides additional information on the peak components separated. RI and conductivity detectors are primarily mass detectors with the peak areas (or heights) proportional to the amount of material in each peak. The fluorometer reacts to specific materials that absorb UV light and reemit it at higher wavelengths. The UV detectors are the most versatile HPLC detectors. Variable detectors can be set to see compounds that absorb only at specific wavelengths, whereas diode-array UV detectors can detect compounds that absorb UV light anywhere within wavelengths available to the array.

The final hardware component in an LC/MS system is a computer, used for data acquisition and processing. This can be used simply as a stripchart recorder or an integrator, or it can be used as a computer-based system for controlling all system components and acquiring peak data, quantifying the peak areas, determining the molecular weights of the components of each peak, identifying impurities, and comparing fragmentation patterns of peaks to known databases to identify each compound present definitively.

2.2   GRADIENT VERSUS ISOCRATIC SYSTEMS

So far we have described only a very simple HPLC, an inexpensive single-pump isocratic system capable of pumping an unchanging mobile phase. Changing the solvent in the reservoir can produce step gradients of solvent for washing out or eluting late-running peaks. For many dedicated applications, this may be all that is needed.

A more complicated system is required for complex solvent separations and for methods development. Solvent gradient chromatography allows separation of complex mixtures of compounds that are poorly resolved under isocratic conditions. Gradient chromatographs produce reproducible, continuously changing mobile-phase composition to the material on the column. The first method of doing this is to add a second solvent pump and reservoir, a mixer, and a pump flow controller to speed one pump while slowing the other pump, to produce a dual-pump gradient system as shown in Figure 2.5. Mixing of the solvents is done on the high-pressure side of the pumping, so any dissolved gases that might be released by the heat of mixing are forced back into solution until after they exit the detector flow cell.

The second common gradient system uses only a single pump but adds multiple reservoirs, a programmable switching valve to connect them, a mixer, and a controller that controls pump flow and the switching valve (Figure 2.6). The advantage of this system is that it is less expensive, since only a single pump is required and more than two mobile phases are available for producing gradients, for methods development, and for automated column washout.

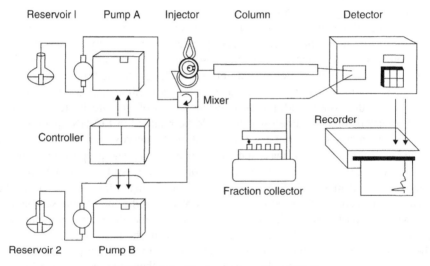

FIGURE 2.5 Dual-pump gradient HPLC.

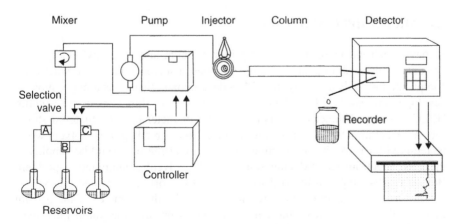

FIGURE 2.6 Mixing valve gradient HPLC.

The switching valve gradient system would be the ideal choice in all cases because of the price and performance advantages if all other factors were equal. However, the gradients produced by this system are not as precise or reproducible as those created by dual-pump gradient systems. The switching valve gradient system also required degassing of the reservoir solvents with helium. The heat of solvent mixing without degassing in the switching valve–mixer combination would cause air to be pulled out of solution in the pump head, resulting in cavitation and vapor-lock blocking of solvent delivery to the column. The oxygen–nitrogen mixture of air forms large bubbles in the solvent that stick to pump surfaces when the pump piston pulls solvents in through the inlet valve. Helium forms small, nonsticky bubbles that are forced back into solution on the compression part of the piston stroke.

Most switching valve systems provide three and four solvent reservoirs, but methods development of three and four solvent gradients are so complex that you rarely see more than a two-solvent gradient with possibly a constant level of another solvent added from a third reservoir. The primary uses of the third and fourth solvents is for automated column washing and method scouting, both of which would probably be eliminated before a method was used in LC/MS analysis. A dial-a-mix addition of constant levels of volatile buffer from this third reservoir might be one use that would be retained in LC/MS operation.

Gradient methods involving solvents with wide polarity and volatility ranges provided problems for the mass spectrometer interface. The purpose of an interface is twofold. It must remove as much solvent as possible without losing the analyate, and it must volatilize and ionize its components for submission to the analyzer. Solvent may not be removed completely if the mobile phase contains too wide a polarity change during the course of the separation, and they may suppress ionization in the interface and ion detection in the mass spectrometer. For the same reason, gradients involving nonvolatile buffers or ion-pairing reagents must be modified using volatile equivalents before they are approved for LC/MS use. This is discussed in more detail in Chapter 5.

2.3   MICRO HPLC SYSTEMS

A mass spectrometer operates at very high vacuum and needs extremely small amounts of material for analysis. Large volumes of solvent tend to overwhelm and complicate the analysis. Early in LC/MS development there was a major move to develop microflow HPLC systems that would use very thin, very high-resolution HPLC columns. Thinner columns and smaller-diameter packing material should increase resolution by decreasing intracolumn band spreading. Less solvent would be required for peak elution, so more concentrated solutions could be supplied to the mass spectrometer, reducing somewhat the solvent contamination problem. But nothing is simple in the real world and a trade-off can turn and bite you.

The smaller the column diameter and the smaller the diameter of the packing material, the higher will be the column backpressure that the pump must overcome. The packing material is held in the column by the fritted filter in the end cap of the column. The pores of the outlet frit must be smaller than the diameter of the packing material or the packing will wash out through the frit. But the inlet frit also acts as a filter for the sample/mobile phase, which means that samples and solvents must both be filtered with finer filters before being sent to the column. The finer the frits, the more they contribute to column backpressure.

In addition, the pumps themselves have to be changed. Piston pumps use stepping motor drives with perhaps 3200 steps per revolution to reduce pulsing and gain precise control over delivery. If the piston displacement is 100 μL/stroke, 10 stokes/min are need to produce 1 mL/min flow rate. At 100 μL/min flow in a microsystem, delivery becomes 1 stroke/min, but if the pump is to be used to create a gradient from 0 to 100%, it must deliver 0.01 stroke/min at 1% delivery. Opening and closing of check valves at these flows is problematic at best. Precise microflow gradient work on piston pumps is very difficult.

Because of this problem, manufacturers have returned to a very old idea in HPLC and have resurrected the syringe pump. A syringe pump is like the cylinder and piston in an automobile engine. The piston is pulled back, drawing in mobile phase through an inlet check valve, and then driven forward to deliver solvent out to the injector through an outlet check valve (Figure 2.7).

The volume of the cylinder is made large enough to provide sufficient solvent for the entire run, and then the cylinder is refilled for the next run. If you want gradients, you simply need a two-cylinder syringe pump and an impinging mixer. Cylinder walls and piston seals have to resist pressures up to 10,000 psi, but flow rates of 10 to 100 μL/min and microcolumn run times of 3 to 15 minutes allow cylinder volumes to be kept reasonably small. Syringe pumps disappeared from conventional HPLC systems in the late 1970s because cylinder volumes had to be very large to allow needed run times to run larger-diameter columns. The cost of cylinder wall pressure armor and replacement seals, solvent compressible changes in flow rate, and the danger of irreparable cylinder scratching made these

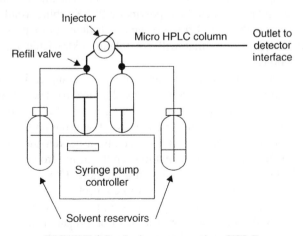

FIGURE 2.7 Syringe-pump micro HPLC.

systems noncompetitive. They did give nice, smooth mobile-phase flow-though. Most microsystems designed for microliter flow rates use syringe pumps or piston pumps designed with short-stroke, very low-solvent-displacement pistons.

It is not just enough to modify the pumping flow rates for a microsystem. Dead volumes in system tubing must be kept to an absolute minimum using 0.005-in.-ID (inside diameter) tubing. Ideally, the injector would screw directly between the pump and the column head, and the column outlet would fasten directly to the detector flow cell. Injector loops and detector flow cell volume must also be reduced. Special microflow injectors are made with internal loops capable of delivering microliter injections. Micro detector flow cells that have a 0.25-μL illuminated volume instead of the standard 20-μL volume are available for secondary detectors.

2.4 HPLC TUBING AND FITTINGS

This section may seem trivial, but extracolumn volumes are critical to optimum HPLC performance. Dead volume in tubing, flow cells, and diverter valves from the injector to the column head can lead to band broadening. Dead volumes from column outlet to the detector can destroy a perfectly good column separation and remix separated elution bands. Inlet tubing is much more sensitive to room-temperature fluctuations than is the column itself. These effects can cause baseline drifting and cycling and are easily overcome by snapping rubber tubing split lengthwise over the inlet tubing to provide an insulating dead space.

Most of the stainless steel tubing used for these critical runs is 0.009-in. internal ID, called *ten-thousands tubing*. Less critical areas such as flush valve outlets and splitter valves use 0.02- and 0.04-in. tubing, and it is important to separate these tubing types so that the wrong type of tubing is not picked up by mistake. Special 0.005-in. stainless steel tubing is available to use with microinjectors and columns. The larger sizes—0.010, 0.020, and 0.040 in.—can be cut with a tubing cutter and polished with a flat file. The finest tubing should be purchased precut and prewashed. Inert polymeric PEEK tubing usable to 4000 psi is available for use with inert HPLC systems with titanium or polymeric wetted surfaces used in applications that will not tolerate extractable metals, such as enzyme purifications.

Tubing is connected to system components using compression fittings made up of a ferrule-and-screw arrangement (Figure 2.8a and b). The screw and ferrule are placed on the tubing line, placed in the connection on the component in which the line will be used, and the ferrule is compressed onto the line by tightening the screw. Try not to overtighten the fitting! Usually, a quarter-turn with a wrench after the fitting is fingertight is sufficient to keep the fitting from leaking. Test it at pressure; if it leaks along the line, tighten it a bit more. Other fitting components are zero-dead-volume unions (Figure 2.8c) to connect two pieces of tubing with fittings and three-way diverter unions used in splitters and flush valves.

Fittings to be used with unions will need to be prepared in situ if the union is to remain a zero-dead-volume connector. An in-line splitter or diverter uses

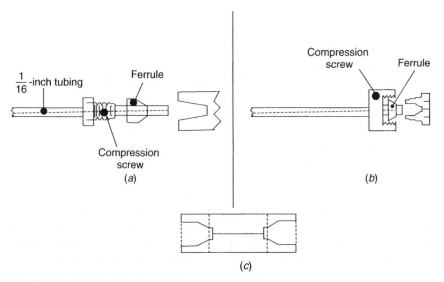

FIGURE 2.8 HPLC compression fittings: (*a*) male fitting; (*b*) female fitting; (*c*) zero-dead-volume union.

0.010-in. tubing in the flow path and larger tubing, usually 0.02-in. tubing, in the diversion path. The fourfold increase in cross-sectional area diverts much of the flow away from the in-line path. For instance, a solvent splitter can be used to send part of the column effluent to a secondary detector while the main flow is sent to an ionizing interface for a mass spectrometer.

One of the most useful applications for tubing, unions, and fittings is to prepare a column blank (Figure 2.9). HPLC systems get dirty over time and must be

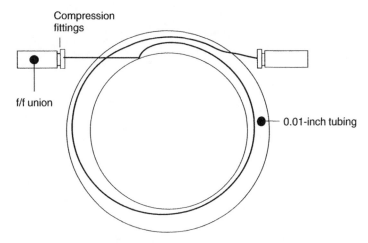

FIGURE 2.9 Column blank.

cleaned. Buffers accumulate on check valves and plungers; sample precipitates in the injector line; and secondary detector flow cells become fogged by effluent materials. Most wetted surfaces of an HPLC system, except the column and mass spectrometer, can be cleaned with water, organic solvents, and 6 N nitric acid. Note that columns are an exception: **Columns never should be washed with nitric acid!** They can be washed with certain organic solvents if buffer is first washed out with water. This may seem obvious and trivial, but I have seen many columns plugged with precipitated buffer, on two occasions I have seen columns ruined with 6 N nitric acid, and on one memorable occasion I saw a $1000 silica protein purification column totally dissolved with Trisma base. To avoid these problems I advocate use of a column bridge. This 5-ft coil of 0.010-in. tubing is equipped with a compression fitting and unions and is used to replace the HPLC column. The column is washed out with water, removed, capped, and set aside. It is then replaced with a column bridge and flow is diverted away from sensitive detectors, such as conductivity, electrochemical, or mass spectrometers. The remaining HPLC system can now be washed with organic solvents such as methanol, acetonitrile, or dimethyl sulfoxide, then with water, and finally with 6 N nitric acid. I usually recommend that the system with the column bridge be washed next with water for 2 hours at 1 mL/min and then over a weekend at 0.1 mL/min before removing the bridge and replacing it with the column. This treatment once a month will prevent a multitude of check valve and injector problems and save a lab many visits from an instrument repairperson.

3

THE HPLC COLUMN AND SEPARATION MODES

The HPLC column is the heart of the separation. All the hardware we have discussed so far is there simple to support getting the sample onto the column, where it can be separated as much as possible into individual bands. These compounds of interest can then be passed to the detector(s) for detection and analysis.

3.1 COLUMN CONSTRUCTION

The HPLC column is a heavy-walled stainless steel tube equipped with inlet and outlet fittings that is pressure packed with fine-diameter packing material suspended in a mobile phase (Figure 3.1). The inlet and outlet column fittings are made up of three parts: a female fitting compressed onto the column, a porous fritted filter sitting within the female fitting directly on top of the column head, and a male end cap drilled to accept a compression fitting connecting a line from the injector or going out to the detector interface. Figure 3.2 shows details of an inlet column fitting.

The porous frit inside the inlet column fitting acts as the last filter before the packing material. In the outlet column fitting it serves as the column bed support and keeps the packing material from being blown out of the column. Preparing the compression fitting on the injector outlet line in the hole on the inlet column fitting allows the end of the transfer line to butt directly on the porous filter for a zero-dead-volume connection to prevent sample dilution by mobile phase. Preparing the column outlet line compression fitting in the hole in

LC/MS: A Practical User's Guide, by Marvin C. McMaster
Copyright © 2005 John Wiley & Sons, Inc.

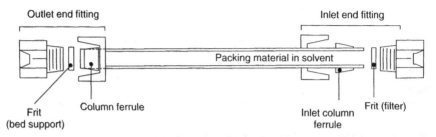

FIGURE 3.1 The HPLC column.

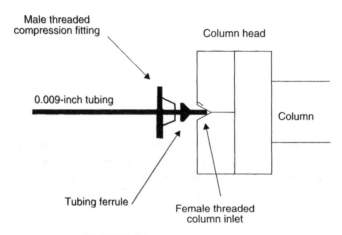

FIGURE 3.2 Inlet column fitting.

the outlet end cap avoids a mixing volume after the separation that could remix separated bands. This may sound trivial, but I have seen good separations ruined when someone used a tubing line with compression fittings prepared on other columns. Peaks sharpened and peak heights jumped as soon as proper tubing fittings were prepared. Minimizing extracolumn volumes from the injector to the detector is critically important in achieving optimum separation, especially when using 3-μm packing and microcolumns.

The metals in stainless steel tubing can cause problems in the separation of biological samples such as enzymes and when using high salt concentrations in the separation. A variety of glass-lined and externally supported polymeric columns have appeared to address this problem, but they not widely used for separations. Inert HPLC systems using titanium or polymeric wetted surfaces should be used to take full advantage of these columns. HPLC columns are packed by removing the inlet end cap and filter, connecting the column to a reservoir of packing material slurry suspended in a working liquid, and pumping it into the column by a high-pressure pump at greater than 10,000 psi. Slurry packers are available, but since high efficiency and reproducible results are critically important, this job should be left to professional column manufacturers. Column diameters and lengths vary with the diameter of the packing materials and the column application. The column industry has standardized on a 4.6-mm-ID column either 250 or 150 mm long for most separations using 5- or 10-μm packing. Microflow columns are usually 2.1 or 1.0 mm ID. Short columns of 50 and 75 mm length packed with higher-efficiency 3-μm packing are used for fast HPLC scouting and for some clinical LC/MS. They are used to trade off the efficiency of using a longer column for faster run times. Separation times can be decreased from 30 minutes to 1 to 2 minutes for simple isocratic runs.

3.2   COLUMN PACKING MATERIALS

Selecting the correct packing material for a column and the correct mobile phase for the separation controls the separation that will be achieved. The most commonly used silica (Si) column is called a *C18 column* or *ODS* (octyldecyl–silica) *column*. It is named for the nonpolar organic phase chemically bonded to the underlying silica surface. Probably 80% of all HPLC separations are carried out on C18 columns or the equivalent. The first analytical HPLC columns were packed with finely ground porous silica with an average diameter of 10 μm. Larger-diameter particles had too wide a difference in the path that a molecule could follow through its pores, leading to band spreading. Irregularly shaped microporous packing exhibited a more consistent pore path and sharp bands in the column. Modern HPLC packing uses spherical particles that pack tighter, have a more consistent particle size and higher efficiencies, and are less bothered by the channeling and voiding problems seen in irregular packing. Free, unbound silanol sites act as ion exchangers and promote hydrolytic dissolving of silica; end capping with trimethylchlorosilane ties up most of these sites, increases column life, and improves the separation of basic compounds. Second-generation type II silica columns are available with more consistent pore sizes and with metal contaminates removed. A bridged organosilica coating provides a surface coating that makes hybrid silica columns much more resistant to dissolving at higher pH values and able to retain their separation characteristics for a much longer time (Figure 3.3).

Particle diameters have moved from 10 μm to 5 μm and finally, to 3 μm. The smaller-diameter particles have a more consistent resolving path, they pack tighter, and are more efficient. Extraparticle volumes are mixing volumes that reduce efficiency. But as size decreases, backpressure increases, extracolumn tubing volumes must be very carefully controlled, and the columns become susceptible to contamination. Support and inlet filters must be of finer porosity and thus are more easily plugged. For most modern analytical work, the 5-μm packing material is the size of choice, but 3-μm packing is used in short columns for rapid assay work and in smaller-diameter columns for high-resolution microflow applications.

Columns based on nonsilica supports have begun to appear in the last few years and have been accepted because of their advantages, even though most of the

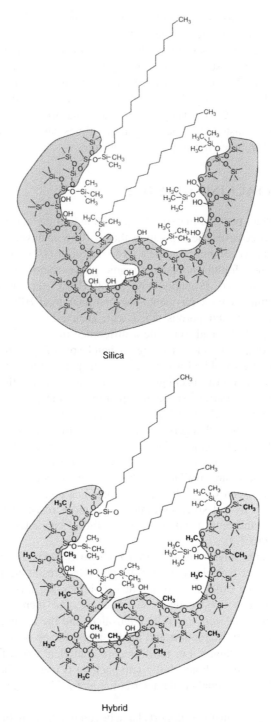

FIGURE 3.3 Silica and hybrid: bonded and endcapped. (Courtesy of Waters Associates.)

separations recorded in the literature were run on silica columns. Silica has three major problems. It dissolves at pH above 8.0, at elevated temperature in aqueous mobile phase, and at high salt concentrations. Most separations are carried out on silica with bonded organic phases coupled to the support surface through a Si–O–Si linkage that hydrolyzes below a pH value of 2.0. This means that the column must be run within the pH range 2.5 to 7.5 at ambient temperature, with salt solutions of 100 mM and less. Going outside these conditions will shorten column life dramatically. If bonded phases bleed off or silica dissolves, column efficiency drops and detectors and interfaces become contaminated.

To avoid these problems, heavily cross-linked polymeric (poly) columns with organic bonded phases began to appear. They are inert from pH 1 to 11, but they show a wide separation difference from silica-based columns and have suffered from slow material transfer into pores, leading to poor loading characteristics and lower efficiency. Elevated temperatures can improve penetration of the pores but also lead to particle swelling, which counters transfer. Poly columns tend to collapse under elevated pressure. Although this has been avoided by creating particles with a high degree of polymer cross-linking, the increased rigidity limits operating pressure and flow rates. Poly columns also differ from silica-based columns in separating character, due to a lack of ionizable surface molecules. Silica at low pH values loses a proton to form anionic Si–O$^-$ moieties, giving a bonded-phase silica column some anionic as well as nonpolar organic column characteristics. This mixed-mode separation is not available in a polymeric column (Figure 3.4a).

The next type of column packing to appear was based on porous zirconium. Zirconium columns come in a variety of particle sizes and nonpolar organic and ion-exchange coatings. They are stable from pH 1 to 12 and from ambient to 200°C. Like silica columns, they add cation-exchange (Lewis acid) effects at low pH to their nonpolar retention character. At high pH values they add anion exchange (Lewis base) and act as metal chelators (Brønstead acids) with an affinity for the free electron pairs on compounds such as amines (Figure 3.4b).

3.3 NORMAL-PHASE COLUMNS

Untreated silica columns are referred to as normal-phase columns because they were the first type of column available for HPLC. They are normally packed in nonhydrated form for use with nonpolar mobile phases. The more polar components of the partition sample in the mobile phase are most tightly retained; the more nonpolar compounds wash out first. These columns have many advantages for use in LC/MS. They resolve positional and structural isomers and show selectivity for compounds with varying numbers of double bonds and aromatic groups. They are usually run in organic solvents that are volatile and more easily removed in the LC interface than are aqueous mobile phases. Silica columns can be used with aqueous acid mobile phases for resolving mixtures of charged compounds such as phospholipids, without the problem of bonded-phase bleed. It is

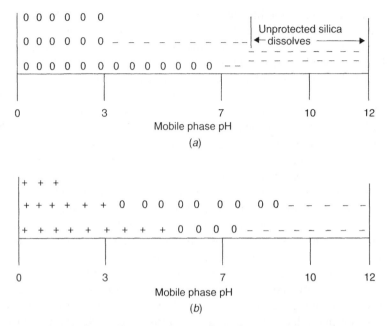

FIGURE 3.4   Silica (a) and zirconium (b) column charge states.

important to understand, however, that once these columns have been exposed
to aqueous mobile phase, they cannot be restored to their original state without removing the packing and baking it at 300°C. They can be washed with
dilute acid, then water, and then with compatible nonpolar solvents back to a
hydrated silica that provides reproducible separations. But these separations will
be different from those obtained with the original nonhydrated column.

3.4   OTHER BONDED-PHASE SILICA COLUMNS

The octyldecyl–or C18–silica column, the first of the bonded-phase columns,
has given HPLC much of its versatility. The octyldecyl side chains are linked to
the silica support by Si–O–Si links. These reverse-phase columns with bonded
organic phases can be run in less expensive aqueous solvents that are easier
to dispose of and less hazardous to use. Other bonded-phase columns quickly
followed the C18 but never replaced it in popularity (Table 3.1). Some 80% of
all separations made on silica columns are run on octyldecyl–silica.

The newest generation of bonded-phase silica columns are called *hybrid silica
columns*. The silica surface of a particle is modified with a cross-linked tightly
adhering silica–organic phase that interacts with an adjacent silanol group and
coats the surface. The remaining free silanols are then reacted with the appropriate
chloroalkyl silane reagent to produce the bonded phase. These bridged organosil-
icone supports are much more resistant to elevated pH and can be used for the

TABLE 3.1   Silica Bonded-Phase Columns

Column	Phase	Solvents	Application
C18	Octyldecyl	AN, MeOH, H_2O	General nonpolars
C8	Octyl	AN, MeOH, H_2O	General nonpolars
Phenyl	Styryl	AN, MeOH, H_2O	Fatty acids, double bonds
Cyano	Cyanopropyl	AN, MeOH, THF, H_2O	Ketone, aldehydes
Amino	Aminopropyl	H_2O, AN, MeOH, THF, $CHCl_3$,, CH_2Cl_2	Sugars, anions
Diol	Dihydroxyhexyl	AN, MeOH, THF, H_2O	Proteins
SAX	Aromatic	Salt buffers	Anions
	Quaternary amine	AN, MeOH, H_2O	
SCX	Aromatic	Salt buffers	Cations
	Sulfonic acid	AN, MeOH, H_2O	
DEAE	Alkyl ether	Salt buffers	Proteins, anions
	Ethyl 2° amine	AN, MeOH, H_2O	
CM	Alkyl ether	Salt buffers	Proteins, cations
	Acetic acid	AN, MeOH, H_2O	
SI	(none)	Hexane, methylene	Polar organics, positional isomers
	Silanols	Chloride, chloroform	

chromatography of basic compounds. They also show increased bonded-phase stability at low pH.

The octyl–silica column has C8 side chains attached to silica through Si–O–Si linkages. The shorter chains allow the nonpolar components of the mobile phase to approach closer to the polar support, and they are not as tightly held as they are on a C18–silica column. It takes less nonpolar solvent in the mobile phase to elute nonpolar components off the column in the same time. A phenyl–silica column has a bonded styryl function bonded to the silica and shows a preference for, and resolves, aromatic compounds and compounds with a varying number of double bonds, such as naturally occurring fatty acids. Aminopropyl–, cyanopropyl–, and diolalkyl–silica columns have intermediate polarity between octyldecyl–and normal-phase silica columns. They can be used with either nonpolar normal-phase solvents or aqueous reverse-phase solvent mixtures. Care must be taken with the aminopropyl–silica columns and other supports with primary and secondary amine groups, since the amino group can oxidize in solvents containing dissolved air over a period of time and lose their separations. Nitrogen- or helium-purged solvents should be used to run, and especially to store, these columns.

These are the most commonly use of the silica-based bonded-phase columns, but a specialized column industry creating column packing for specific functions has grown up. Affinity columns are prepared with an alkyl linker group bound to the silica that can react with an active molecule that can trap and retain only a single class of compounds in the mobile phase. Once trapped, the affinity target can be removed by changing mobile-phase conditions to introduce molecules that force the target off the bonded phase. Generally, these include pH control, salt concentration, or specific molecules that bind tighter than the target. A common use for these columns is for protein and antibody purification. The eluants used often bar them from use in direct-flow LC/MS, but combined with a desalting size-separation column, they could be used with a mass spectrometer. Since these column supports must be prepared specifically for the separation they are designed to make, they are most likely to be found in a production rather than an analytical environment.

Enantiomeric–silica and poly columns are designed to resolve mixtures containing optical isomers. These compounds have at least one asymmetrical center, and a partition separates compounds that are mirror images of each other. The bonded phase of the column contains a resolved optical isomeric functionality that will interact differently with each optical isomer, leading to a difference in portioning rates. One compound of an enantiomeric pair is more tightly retained, and the pair is separated as it travels down the column. These columns are important in resolving optically active pharmaceuticals, sugars, synthetic peptides, and proteins.

3.5   OPTIMIZING REVERSE-PHASE COLUMN USE

Reverse-phase separation is supposed to be a simple partition of the sample between a nonpolar bound phase and a polar mobile phase. The most polar sample components should come off first, with the nonpolar compounds being retained longest on the nonpolar column packing. But nothing in life is that simple. Even in the most heavily end-capped columns, the underlying anionic silica enters the picture and affects the separation that is achieved. Compared to polymer-based columns, this mixed-mode separation actually adds to the resolving power of silica reverse-phase columns, allowing them to make separations that other column supports cannot achieve. As reverse-phase silica columns age, they change their running characteristic, due to hydrolysis of the bonded phase freeing up more and more silica sites. Even if the column is run only between pH 2.5 and 7.5, they can still bind free metals, such as calcium and magnesium ions, from glass reservoirs used for mobile phase and change their resolving character.

There are ways to overcome or adjust to these changes: One is column cleaning, covered in more detail later in this section. When I was an HPLC salesman, I had a customer in Cleveland who cycled his octyldecyl–silica columns through a series of four routine separations. He would start a new column on his most critical separation. When that separation could no longer be achieved, he washed out the column and used the column for the next-most-critical separation. Following this procedure he was able to get a 24-hour-a-day, two-year lifetime from each column. Most of my customers would have been happy with a 1000-injection lifetime for any column.

Buffers are used with reverse-phase columns primarily to control ionization of sample molecules so that they can undergo partition interaction and adhere to the bonded phase. Ionized forms usually tend to be too soluble in the mobile phase and will wash out in the solvent-front first peak. Ionizations of organic acids can be suppressed by buffering two pH units below their pK_a value of around 4.5. Amines have a pK_a value of about 10, and silica columns adjusted to this pH will simply dissolve. Amines can be retained and separated on nonpolar columns by adding an ion-pairing agent such as an organic sulfonic acid to the mobile-phase analysis. Both buffers and ion-pairing reagents create problems for LC/MS analysis if they are nonvolatile. They are not removed in the ionizing interface, causing either ion suppression or contamination of the analyzer signal. Nonvolatile metal salt buffers, such as potassium phosphate, can be replaced with volatile buffers such as ammonium acetate or ammonium formate.

Bonded-phase columns are usually shipped and stored in at least 25% organic solvent to prevent bacterial growth when they are not in use. This storage solvent needs to be washed out when a column is to be used, and the column need to be equilibrated with the mobile phase. Be careful not to switch from organic solvent to buffer or buffer to organic solvent if the buffer will precipitate on contact with the more nonpolar solvent! Wash out the organic solvent with water first and then switch to the buffer solution. Generally, you should wash a column with six column volumes of the mobile phase before you make an injection. Six volumes for a 4.3-mm-ID column 25 cm long is about 18 mL of mobile phase. At that point, although the column is not fully equilibrated, you should be able to make reproducible chromatography injections. True equilibration may take overnight pumping of the mobile phase to achieve and does not greatly improve the separation.

When you are finished with the column, reverse the procedure for storage. Wash out buffer with water, then wash the column with 25% organic and cap the column for storage. Please note that mixes of polar organic solvents with water may have a dramatically higher viscosity than water or the solvent alone and can cause overpressure problems at high flow rates, so watch the pump pressure gauge during the run and drop the flow rate if necessary.

As nonpolar contaminates begin to build up, you will notice that late-running peaks of a check standard mixture begin to take longer to come off or disappear completely from the chromatography. Washing a column is an art unto itself. First, **do not wash the column into the mass spectrometer**. That is what a diverter valve is for, to divert the wash solvent away from the interface and the analyzer. Some detectors can benefit from the wash—but, not the mass spectrometer, please! Even the mass spectrometer repairperson will thank you. Nonpolar materials tend to accumulate on nonpolar columns and can be washed out with nonpolar solvents. Make sure that you first wash out the buffer with water before doing an organic washout. For organic washing of a octyldecyl silica column, use solvents in the order methanol, acetonitrile, dimethylformamide, tetrahydrofuran (THF), methylene chloride, chloroform, and hexane. Never go to a stronger nonpolar solvent if a weaker solvent will do. You should use six column volumes of each solvent. Washing time is costly, so if you have to wash all the way to hexane to remove a nonpolar material, try bridging from water through THF to hexane. Make sure that you use six full column volumes before you switch to hexane. Fortunately, both THF and hexane have low viscosities, so you can jump the flow rate at the nonpolar solvent end.

Metal cation buildup is more subtle than organic contamination. Usually, fine separations in the center of a chromatogram will begin to merge with an adjacent major peak and eventually disappear. They can be restored in many cases by washing out buffer from an octyldecyl silica column, then washing the column with 100 mM sodium oxalate (pH 4.0) for six column volumes, followed by copious washing with water. Do not exceed six column volumes, as oxalate is a metal chelator and will extract iron and other metals from the column wall and wetted surfaces. Do not try to switch to EDTA for this wash, as it will not penetrate the particle pores deeply enough to extract the calcium and magnesium ions.

3.6 SILICA ION-EXCHANGE COLUMNS

The first columns used in high-performance liquid chromatography were ion exchangers bonded to silica. These supports are now available in microparticulate packing. SAX material is a silica anion-exchange column with a quaternary amine at the end of the bonded phase. The SCX column is a silica cation-exchange column with a sulfonic acid group at the end of the bonded phase. These columns require a high concentration of salt and salt gradients to elute the retained ionized material from a column. The wetted surfaces of the pumping system and detector must first be protected from salt corrosions and metal leaching by periodic pacification with nitric acid. Replace the column with a column bridge, divert the flow away from detectors such as the mass spectrometer, and then pump 6 N (20%) nitric acid. After this treatment the system must be washed with copious amounts of water before the column is reinserted.

Strong ion exchangers tend to retain irreversibly very strong ions such as proteins, quaternary amines, and sulfonic acid groups that will deactivate the packing material. Weak ion exchangers with diethylamineethyleneamine (DEAE) and carboxymethyl (CM) functions have been prepared bound through alkyl group to silica. These types of functionalities have long been used to purify proteins and are very useful HPLC packing material that can be washed with lower salt concentrations and by pH control of the bonded function group. Like amino–silica columns, DEAE–silica columns need to be protected from oxidation by dissolved air in the mobile phase. Ion-exchange columns that have to be eluted with salt concentration have little use in direct-flow applications into the mass spectrometer interface because of the problem of getting rid of the nonvolatile eluant salts. Separated fractions from the ion exchanger can be passed through a desalting size-separation column into a mass spectrometer interface with a diverter valve used to discard the salt fraction when it comes off.

Hydrophobic interaction columns are basically bonded nonpolar columns used with ion-exchanger mobile phases. They are started at very high salt concentrations such as 100 mM ammonium sulfate, which helps drive nonpolar compounds onto the stationary phase. A gradient is run to low salt concentrations to elute sample peaks, with the most nonpolar compounds being eluted last. This technique is used most commonly with proteins, and because of the high salt concentrations used would have little application for direct-flow LC/MS unless followed by treatment with a size-separation desalting column.

3.7 SILICA SIZE-SEPARATION COLUMNS

The last major type of silica-based HPLC column uses size as their separating mode. These bonded-phase silica columns have controlled silica pore sizes and are tightly coated to mask the underlying silica surface. They only separate materials within a specific size range, excluding materials too large to enter the pores. The largest compounds are eluded first, followed by the resolved material, and finally, a peak containing everything that penetrates the pores completely. These columns are widely used to purify proteins, polysaccharides, and a variety of polymers from monomeric materials. They work well for LC/MS if the molecular weights of the resolved materials do not exceed the resolving range of the analyzer. A diverter valve should be used to discard the excluded or totally included fractions.

3.8 ZIRCONIUM BONDED-PHASE COLUMNS

Bonded-phase columns based on zirconium rather than silica offer some interesting possibilities for LC/MS. Zirconium bonded-phase columns show an even more complex mixed-mode separation than silica columns. They can be used at a wide pH range and show anionic, cationic, and chelating attraction for amines. Often, an amine buffer and a very strong chelating compound must be added to the mobile phase simply to allow amino compounds to elute from the column.

The zirconium bonded phases for use with mass spectrometry are polymeric polybutadiene or C18 groups bound by a direct carbon-to-zirconium linkage prepared using diazo compounds. Unlike the silicon linkages in silica bonded phase, C–Zr supports are not susceptible to acid hydrolysis and the zirconium does not dissolve at higher pH values. This provides HPLC packing material that can be used from pH 1 to 11 and at elevated temperature. The zirconium columns exhibit nonpolar compound partition characteristics similar to those of octyldecyl–silica columns, but are dramatically different when separating charge molecules such as organic acids and amines. Because their nonpolar selectivity differs from C18–and C8–silica-based columns, they offer a possible solution for problem separations. Like silica, zirconium-based columns show cation-exchange characteristics but add both anion-exchange and metal chelating character. These can be used to change the separating character of the bonded phase or can be suppressed with buffers and chelating agents such as ethylenediaminetetraphosphonic acid (EDTPA).

Sometimes, these differences can be too much of a good thing since most of these columns have to be treated with chelators to keep them from irreversibly retaining cations and amines. Columns are now available with a bonded nonpolar organic phase and covalently bound tetraphosphodiamine chelators ideal for LC/MS. The zirconium columns have no column bleed problem, mobile phase can be buffered with volatile buffers from high to low pH, and they can be run at temperatures as high as 80°C to speed peak resolution and shorten the chromatographic run time.

Zirconium bonded phases can be washed by a three-step procedure: (1) wash with 80% acetonitrile containing 100 mM ammonium hydroxide (pH 10.0) for 50 column volumes; (2) wash with 80% acetonitrile with 100 mM nitric acid for 50 column volumes; and (3) wash with 100% organic, MeOH, acetonitrile, or THF for 20 column volumes. Obviously, this should be done using a diverter valve so that the washes do not go through the mass spectrometer. These column volumes, specified by a column manufacturer, seem a bit excessive.

3.9 POLYMER REVERSE-PHASE COLUMNS

Polymeric bonded-phase columns offer pH stability from pH 1 to 13, do not suffer from the competing mixed-mode separations seen in silica and zirconium bonded-phase packing materials, and can be used with pure organic solvents although intermediate solvents such as THF and dioxane may cause swelling-induced pressure problems. Reverse-phase columns provide true nonpolar resolving characteristics. Bonded-phase polymer columns are also available with anionic and cationic exchange character and with ion exclusion phases.

Check the pressure limits specified by the column manufacturer before using this type of column. Some can tolerate column backpressure to 5000 psi, but the polymer cross-linking that makes this possible also limits mass transfer into the pores and decreases sample loading. These columns cannot be used at elevated temperatures, which cause polymer swelling and pore closing. The only buffering necessary with these columns is to control sample ionization-reducing contaminates, which might interfere with mass spectral analysis. Their main limitation is that they may fail to resolve all the compounds of interest in a mixture.

4

HPLC AND COLUMN MAINTENANCE

HPLC is a very reliable and proven technique, but the hardware and columns are subject to contamination problems. Reliability and reproducibility are greatly improved with a small amount of consistent preventive maintenance and understanding of the limitations that apply to the use of each type of column.

4.1 HPLC MAINTENANCE

By far the most troublesome parts of an HPLC system are pump check valve blockage, injector loop contamination, and column contamination. In 25 years of selling, demonstrating, and troubleshooting HPLC systems and columns, I have found that 90% of all difficulties are caused by column problems, most of which come from contaminated water. The following tasks should be performed daily on a LC/MS system:

1. Inspection for leaks in pump seals, plungers, and check valves
2. Routine syringe and injector cleaning
3. HPLC column washing
4. Interface washout and maintenance
5. Secondary detector flow cell maintenance

4.1.1 Pump Pressure Checking and Pacification

Pressure checking is a diagnostic tool for plunger and check valve performance or, more to the point, nonperformance. Monitoring changes in the pump pressure

LC/MS: A Practical User's Guide, by Marvin C. McMaster

during initial pressurization can provide information on check valve contamination and broken plungers. As noted earlier, most pump plungers are broken when
the pump head is being checked. It is better to use a noninvasive diagnostic.
The first tool needed for diagnosis is a column blank to provide backpressure to
ensure check valve closing (Figure 4.1).

Removing the column simplifies the diagnostic problem since the column is
the major source of problems. This is the equivalent of eating an elephant by
starting on the trunk instead of cooking the entire beast at one time. As noted in
Chapter 2, the column blank is a 5-ft coil of 0,009-in.-ID tubing equipped with
a compression fitting and zero-dead-volume unions on each end. It will provide
about 75 psi of backpressure to a pump running aqueous solvents at 1.0 mL/min.

Place the pump inlet in a 50-mL graduated cylinder of degassed solvent. Prime
the pump to clear air bubbles out of the pump head by cracking the compression
fitting at the head of the outlet check valve to release the air, if necessary.
Connect the column blank to the outlet filling of the pump with 0.009-in. tubing
and connect another small piece of 0.009-in. tubing to the outlet end of the
terminal union and place it into a 10-mL graduated cylinder. Turn the pump
on at 1 mL/min and watch the outlet stream; it should be a fine, steady jet of
liquid. Each plunger stroke should consistently deliver 50 to 100 μL of solvent,
depending on the manufacturer. Watch for periodic change or ceasing of flow at
these increments. If the equipment consistently stops delivering solvent half of the
time, the plunger may be broken or an outlet check valve may be stuck. Watch the
level of the solvent in the inlet graduate. A contaminated inlet check valve will let
solvent run back into the reservoir cylinder, causing the level to rise and fall. Next,

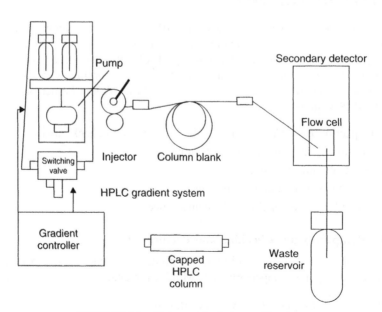

FIGURE 4.1   Column blank for pacification.

empty the outlet graduated cylinder and use a stopwatch for 9 minutes to check if the pump is delivering 9.0 mL. Leaking check valves or broken plungers will deliver much less than the volume expected. Do not get upset at minor deviations from the volume expected. Pumps are not always calibrated for exact delivery, and consistent performance is more important to a chromatograph than accurate delivery. If you have a multiple-head pump, connecting a column bridge directly to each check valve can tell you which pump head is malfunctioning.

Now for the good news: None of this may be necessary if you will take time once a month to pacify your system with 6 N nitric acid after first removing the column and replacing it with a column bridge. This is very important! **Columns do not like nitric acid; it ruins them!** Wash out the column with solvent without the buffer, and then wash it with the storage buffer, cap it, and set it aside. Replace it with the column bridge. **Do not wash nitric acid into a mass spectrometer or its interface!** The interface may be resistant, but you will have to check it with the manufacturer. Make sure that the outlet of the column bridge is connected to the secondary UV detector's flow cell if it is resistant to nitric acid. Most flow cells can be washed with this reagent; in fact, it does a very nice job of cleaning them. But check with the manual or the manufacturer. If nitric acid is not recommended, place the column blank outlet into the reservoir.

Now we are ready to start. *We have the column out*, and we have the column bridge in place. Turn the injector to the load mode so that the loop gets washed. Wash out the system with water at 2 mL/min for 15 minutes and then change the solvent to 20% (6 N) nitric acid and wash into a collection reservoir for 30 minutes. I would suggest that you set up a regular routine of pacification at 4:00 P.M. on the last Friday of each month. Make a habit of it. Use water, nitric acid, then water again. Replace the nitric acid inlet reservoir with 1 L of water, turn the flow rate down to 0.01 mL/min, and wash over the weekend into a clean 1-L flask. Collect fresh effluent when you return on Monday, check its pH to make sure that it has returned to the same pH as that of your HPLC water, and replace the column. Wash out the storage solvent and reequilibrate with your mobile phase.

Pacification will clean out organic and buffer accumulation on pump check valves, plungers, and seals and from the injector loop, tubing, and detector flow cell. I have found that it will improve seal and plunger life and prevent almost all check valve replacement. But it must be done with the column out of the system. I have had students who did not listen to that warning, and their columns were never the same. People will laugh and point at you if they find out that you have pacified a column. Take the column out first, then pacify; save yourself the embarrassment!

4.1.2 Injector and Tubing Clearing

Injectors plug up for one of three reasons: Samples were not filtered before they were injected; samples were made too saturated and precipitate when concentrated at the column head; and solvents containing buffer were switched too rapidly to solvents in which the buffer was not soluble. Samples should always be filtered through filters with pores smaller than the pore size in the column frit at the head of the column. Sample concentration at the column head is a problem usually seen in preparative chromatography but seldom in LC/MS. Buffers containing solvent should always be washed out with six column volumes of water before moving to another organic solvent.

The injector outlet tubing is the point where tubing size decreases to the smallest diameter in the pumping system. This is done to decrease extra-column volumes and prevent band spreading at the column head. The injector end of the tubing leading from the injector to the column is the point most likely to plug. Switching this line end for end and blowing it into a beaker using pump pressure will often unblock it. If this does not clear the line, the plugged portion in the first few millimeters should be cut off, a new ferrule added, and the fitting remade in the injector outlet hole.

4.1.3 Gradient Performance Diagnostics

Solvent gradient chromatography is only as good as the pumps and switching valve that generated the gradient. If is often difficult to tell what went wrong with a system when the chromatography changes. Gradient performance in a two-pump gradient system can be judged by using solvents in one of the inlet reservoirs spiked with a UV-absorbing compound such as acetone. The same technique can be used to uncover a failed switching valve in a solenoid valve–based single-pump gradient system. Set up the system with either a column or a column bridge and place 50:50 acetonitrile/water in all solvent reservoirs. Spike reservoir B to 50 mM with acetone, set the UV wavelength to 230 nM, and start recording the UV baseline.

Switching valve problems can be diagnosed by running 0% B to establish a baseline, then a step gradient to 100% B, and reequilibrate at 0% B. No baseline increase will appear in the chromatogram trace if the solenoid valve for the B solvent has failed. Switching the spiked solvent to reservoir C allows us to check % C in the same manner.

Once valve integrity and function are proven, gradient performance can be studied using the same spiked solvent through a series of 5% step gradients up to 100% and back down again. Watch gradient performance at the beginning and end of each step of the gradient. Slope changes should be sharp and vertical, each step should be smooth, and there should be no overshoot at the high end as the new baseline is being established. A really good gradient should be able to repeat this performance with 1% steps, but most systems will have trouble meeting this test.

4.1.4 Water and Solvent Cleaning

Water used to make HPLC solvent needs to be HPLC-grade water. Nonpolar impurities in the water will accumulate on nonpolar columns, contaminating them and changing the resolving character of the column. Even triple-distilled water can have coeluting nonpolar impurities that will interfere with PTH–amino acid analysis during one day's injections. If HPLC water is not available, purify the best water available through either a spare C18 column or a C18 SFE cartridge column. Be aware that the cartridge column has only limited capacity, about 25 mg of organic, and will treat only 1 L of really good water without regenerative washing with 100% organic solvent. Be sure to prewet the cartridge with methanol, then with water, before using it to clean your mobile-phase water.

4.2 COLUMN MAINTENANCE

Remember that like attracts like. All columns get dirty and need to be washed clean. Silica-based columns dissolve at pH values above 8.0. All columns need to be protected by setting the HPLC pump overpressure value.

4.2.1 Silica Column Cleaning

Column washing should always be diverted away from the mass spectrometer! I always disconnect the column from the mass spectrometer interface and wash the column into a beaker. Diversion of the wash through the interface is not usually a good idea, since even secondary UV detector flow cells can be fouled by column washing.

Silica columns are very rugged. The silica itself can tolerate pressure to 12,000 psi except for very large-pore-size columns such as the TSK6000 protein purification column. Early irregular packing columns had an unfortunate, often fatal, voiding characteristic that caused a channel to form in the column bed during repeated repressurizations. Modern spherical column materials are more rugged. But columns still should be handled with care and run in only one direction, to avoid channeling. When switching solvents, immiscible solvents should be avoided and at least six column volumes of the new solvent should be pumped through the column before you can expect to see reproducible chromatography. Care must be taken not to precipitate buffer and plug the column inlet by going directly to an organic solvent in which the buffer is not soluble. Wash out the buffer solvent with water as a bridge.

The biggest problem seen with nonpolar silica columns such as octyldecyl, octyl, phenyl, and cyano columns are nonpolar impurities accumulating at the column head and changing the separation by holding up the more nonpolar components of the injection mixture. These can usually be washed out with acetonitrile if buffers are first removed with a water washout. If this does not restore separation of a standard check mixture, you may have to do more dramatic washing. A three-step wash going from acetonitrile to tetrahydrofuran to hexane and back down will remove all nonpolar impurities.

Protein separation columns are large-pore-size bonded-phase silica columns, generally coated with a diol-type bonded phase. They can be washed like a normal reverse-phase column, but with alertness to overpressure problems caused by viscosity changes due to solvent mixing. The very large-pore-size columns used for separating large protein, antibodies, and DNA restriction fragments are like Christmas tree ornaments and break easily. I have seen the packing in a TSK6000 column crushed to a pancake in the bottom of the column casing because it was shipped capped across the northern Pacific in the winter in 25% methanol/water. Solvent freezing inside the capped column put more pressure on the packing than it could tolerate. Read column brochures that come with a column and set the pump pressure limits that are recommended. Also be aware that there is a difference between Tris phosphate 7.0 used to dissolve proteins and Tris base. A 100 mM Trisma base solution has a pH of about 10.0 in water. I had a customer (who claimed to be a biochemist and should have know better) who pumped this solvent though a 160-cm loaner silica-based protein separation column and returned an empty column casing to me—a very expensive mistake that you will want to avoid.

In the face of extremely tightly held contaminates, I have washed C18 and TSK protein separation columns successfully with 20% dimethyl sulfoxide or dimethylformamide solutions in methanol. Be aware that these solvents have a very high viscosity and have to be run at much slower flow rates to keep from overpressurizing the column.

A secondary problem with silica columns is the accumulation of metal ions on the acidic, free silanol groups. These ion exchangers can be freed by washing the column with water and then with six column volumes of 100 mM sodium oxalate (pH 4.0). This chelator will strip off calcium and magnesium ions that have accumulated from water stored in glass bottles. It will also attack the iron in the walls of the stainless steel column in which the silica is packed if you greatly exceed the washing volume, so proceed with care. I usually assume that a 4.2 × 300 mm column will have about 3 mL of column void volume, and that number seems to work for calculating washing-out volumes.

Polar silica columns such as normal-phase silica and amino columns accumulate polar contaminates and can be washed with polar solvents. Be aware that unless a silica column is run in anhydrous solvents, it will change its initial separating character and will have a permanently different separating nature. As they come out of the box, these columns have an anhydride surface, but around water containing solvents, they convert to a hydrated silica surface. They can consistently be washed back to the hydrated form, but to restore the original surface, the silica would need to be dumped out of the column, dried in an oven, and then repacked. Washing with chloroform is usually enough to remove most polar compounds. If this is not sufficient, do a three-step wash with hexane, tetrahydrofuran, 10 mM acetic acid, and back down.

Primary and secondary amine columns such as amino and DEAE columns will air-oxidize and lose their separating ability. They need to be run and stored in solvents degassed with nitrogen or helium.

Ion-exchange silica columns are bonded-phase columns with negatively or positively charged functionality. They accumulate nonpolar compounds that can be removed by washing with organic solvents, but their most important contaminates are compounds with a strong functional group that has an opposite charge. These adhering compounds must be removed by controlling the pH to neutralize one of the charged groups, either on the bonded phase or from the compound attracted to it, or by displacement using a counterion, such as high concentrations of sodium chloride. This limits column life since salt both attacks the underlying silica and corrodes the column wall.

4.2.2   Zirconium Column Cleanup

The main advantage of the zirconium family of columns is their stability from pH 1 to 11 and at high temperatures. They are Brønstead acids and bases, which ionize at both high and low pH, unlike silica, which ionizes at low pH but dissolves at pH > 8.0. Zirconium also acts as a chelator for Lewis acids, so columns recommended for LC/MS generally come with a covalently attached chelating agent such as ethylenediaminetetraphosphonic acid (EDTPA) to tie up the Lewis base sites on the zirconium surface. This bound chelator allows separation similar to that of silica at acid pH, but also allows amines to be run at high pH. Like silica-based columns, these bonded-phase columns accumulate nonpolar organics that will change their running characteristics unless removed periodically. The manufacturer recommends washing zirconium-based columns for mass spectroscopy first at high pH in 80% acetonitrile with 100 mM ammonium hydroxide (pH 10.0), followed by water, flushing at low pH with 20% acetonitrile containing 0.1 M nitric acid, water again, and finally, an organic solvent such as methanol, acetonitrile, or tetrahydrofuran. These columns require 20 to 30% less organic solvent for eluting equivalent nonpolar compounds than is required on a C18 silica-based column. Chelators in the mobile phase that might interfere with the blocking EDTPA molecule should be avoided.

4.2.3   Polymeric Column Protection

Polymeric columns are a very mixed bag of physical structures and bonded phases. Traditional polymeric columns used for size and carbohydrate separations are physically fragile supports that crush easily and must be run under carefully controlled pressures. Columns were sometimes destroyed simply by starting up with cold solvent if the operator had failed to set pressure protection. Warming the solvent reservoir in a warm-water bath was sufficient to reduce viscosity and allow column operation. Modern polymeric columns are much heavier cross-linked materials and resist moderate pressure, but the recommended pressure setting still needs to be observed. Nonpolar polymeric columns are usually easier to wash than silica columns because they do not have the secondary ion-exchange character. Aqueous acetonitrile or tetrahydrofuran is sufficiently nonpolar to remove most contaminates, and both have a lower viscosity than that of water mixtures with methanol. Heating the solvent reservoir can shorten washing times by reducing viscosity and increasing mass transfer out of the column pores. Be careful not to exceed solvent boiling points or the column may become vapor locked or voided. Table B.1 in Appendix B lists the physical characteristics of common HPLC solvents. Strong ion exchangers on polymeric supports can be treated with high salt concentrations at low or high pH, and at high temperatures, without attacking the polymeric surface. But be aware that high salt concentrations will still attack the iron in the stainless steel column casing, so keep wash volumes to six column volumes and follow with a water wash to neutralize.

5

SAMPLE PREPARATION AND SEPARATIONS DEVELOPMENT

Sample purification begins by selecting the correct column type, then continues by correctly preparing the mobile phase and equilibrating the column for the sample injection. Next, the sample must be cleaned up and filtered before injection. Finally, the separating conditions must be selected and optimized to give the desired separation before passing the sample to the mass detector for analysis and quantitation. I am a great believer in developing the separation method using a detector other than the mass spectrometer, translating the separation by switching to volatile buffers and ion-pairing reagents if the method was not designed with LC/MS in mind, and optimizing solvent removal/ionizing conditions in the interface before passing any sample into the mass spectrometer's analyzer.

5.1 MOBILE-PHASE PREPARATION

In preparing mobile phases, it is important to start with HPLC-grade solvents and water. I have seen a variety of off-grade solvents used in separations, and the results have never been satisfactory. Triple-distilled water may be fine for enzyme reactions, but it can result in nonpolar compounds accumulating at the head of a column and destroying a separation in less than 4 hours. Eventually, these compounds will bleed into the mass spectrometer and mess up an analysis.

If you are not sure about the quality of the HPLC water, swirl it and inspect for floaters, which indicate bacterial growth. If it is clear, pump it down a C18–silica column at 1 mL/min for 30 minutes to accumulate any organics in the water on

LC/MS: A Practical User's Guide, by Marvin C. McMaster
Copyright © 2005 John Wiley & Sons, Inc.

the column, and then run a gradient from 0 to 100% acetonitrile in 20 minutes. Retained organics should appear as peaks somewhere between 20 and 100% acetonitrile.

Once mobile phase has been prepared, it needs to be filtered through compatible 0.54- or 0.22-μm filters and degassed, to remove air before being pumped into the column. If mobile phase is not degassed, it will cause the pump head to cavitate and stall. Be careful about the type of filter you chose; some are designed for organic solvents and some for aqueous solutions. Even those designed for aqueous solvents often have to be wetted with methanol or acetonitrile before they will pass the mobile phase and degass it. Pour the filtered solvents carefully down the side of the mobile-phase reservoir to avoid remixing air into the solvent.

Tubing line filters are mostly sinkers. They contain fairly large pore filters and will plug if you forget to filter solvents or mix incompatible organics with buffers that can precipitate. Insert the pump inlet tubing line filters into the solvent reservoirs and prime the pump by pumping a mixture containing at least 20% organic out through the injection waste line. Pump heads wetted with water or aqueous buffers can trap air bubbles. This problem is reduced with some organic in the pumping solvent, but air bubbles may still have to be removed. Try lightly tapping the outlet check valve assembly with a compression-fitting wrench or combine this with cracking open the outlet compression fitting for a moment while solvent, is flowing. You will get an upwelling of solvent, and as you tap you should see tiny air bubbles. Check the flow rate by timing solvent collection into a graduated cylinder. Connect a column tubing bridge to the injector outlet to aid in closing the pump outlet check valve and ensure that the pump is providing constant flow. Run a gradient from 20% organic solvent up to 100% and back down to make sure that both reservoir lines are free of bubbles. Connect a UV detector, check the detector outlet for fine bubbles, and watch the detector baseline for spikes. A tubing bridge made from 5 ft of 0.01-in. tubing will provide plenty of backpressure to close the outlet check valves.

Filter and degas solvents before using them even if you are using a switching valve gradient system. Your system should provide a method for additional degassing of the solvents by helium purging. This needs to be running whenever solvent is being pumped through the system.

Once solvent is flowing and bubbles have been removed, reduce the flow, return to the starting mobile phase, connect the column inlet line to the column head fitting with solvent flowing, and connect the solvent outlet line to the mass spectrometer detector interface with the diverter valve open so that sample does not go to the mass spectrometer. Flush at least six column volumes of initial mobile phase through the column before closing the diverter valve and sending sample to the analyzer.

5.2 MOBILE-PHASE PH CONTROL USING BUFFERS

It is important to remember the cation-exchange character of the underlying silica support that can affect the separation of charged compounds. Adding buffers to the mobile phase can be used to control ionization of both sample compounds and the underlying silanols. Column packing that has had free silanols end-capped with chlorotrimethylsilane can reduce this ion-exchange effect, but it often returns as a column ages by acid-catalyzed hydrolysis. For LC/MS it is important to use volatile buffers that can be removed in the ionizing interface. Ionized compounds are usually not retained well on nonpolar columns. They can be induced to partition onto the column by adjusting the pH of the mobile phase with buffers supplying or taking up hydrogen ions, forcing them into their un-ionized form.

To separate organic acids, select a buffer with maximum buffer power around a pH of 2.5, two pH units below the pK_a value of the acids. Organic amines are ionized below a pH of 7.5 and tail badly, giving broad peaks when trying to separate them on silica-based columns. Buffering alone cannot be used to induce them to partition onto nonpolar column supports unless they are very large molecules. Amines can be induced to separate on a C18–silica column by adding an alkylsulfonic acid to the mobile phase, resulting in an ion-pairing reagent. These compounds are thought to form an ion pair, with the amine in solution masking the ionized portion of both the amine and the ion-pairing reagent. A second theory explaining the action of the ion pair is that its nonpolar end interacts with the C18 brush phase to form a nonbonded ion exchanger, which then interacts with the charged amine. Equilibrating the column with 5 mM of hexanesulfonic acid added to the mobile phase will allow amines to be retained and separated. Each amine is retained longer if octanesulfonic acid is used as the ion-pairing reagent. Ion-pairing reagents may cause a problem in a mass spectrometer because they are not very volatile. Some work has been done with ammonium propylsulfonate, but if you can get along without using ion pairing, do so and avoid the problem.

Zirconium columns have the advantage of being stable from pH 1 to 11. Amines on these columns can be buffering above their pK_a values and separated as nonpolar compounds giving sharp peaks. It is important in this case to choose columns designed for mass spectroscopy using covalent bound chelating compounds to deactivate the Lewis acid sites of the zirconium substrate, which tend to grab and hold free amines.

Nonvolatile buffers, ion-pairing reagents, and chelators will obviously cause major problems in LC/MS analysis. They can cause ion quenching and contamination of the interface, and if they can be ionized, they are added to signal contamination in the mass spectrometer analyzer. Inorganic salts should be avoided at all times, small organic salts should be substituted and concentrations kept as small as possible, and columns treated with strong chelating compounds such as ethylenediaminetetraphosphonic acid should be used to deactivate the Lewis acid sites of zirconium columns to avoid column bleed. Manufacturers of zirconium-based columns for LC/MS also recommend avoiding the use of fluoride-containing buffers and metal chelating compounds or proteins in addition to nonvolatile salts. Table B.2 is a list of buffers that have been used in LC/MS separations.

5.3   SAMPLE PREPARATION

While a column is equilibrating with mobile phase containing the proper volatile reagents, we can move on to preparing a sample for injection. Samples need to be dissolved in the starting mobile phase and filtered. Try brief ultrasonication if the sample does not dissolve immediately. In a few cases it may be necessary to dissolve the sample in mobile phase containing a higher percentage of the stronger solvent B to speed solution. As long as only analytical-size injections are made, this will not affect the separation, and in fact, will provide a refractive index upset equal to the column void volume, which can be used in calculating the retention times of sample components. Like solvents, samples should be filtered with 0.54- or 0.22-μm filters before injection. In an earlier book, *HPLC: A Practical User's Guide*, I suggested using a clinical centrifuge to spin paired samples in conical minitubes at 750g for a couple of minutes as a method of precipitating particulates before pulling up the sample solution for injection. This technique will work but it will not guarantee that your column filter will be protected from plugging particles, especially if you are using 3-μm columns. Recently, a variety of low-dead-volume solvent-compatible filters in Luer-type tips has become available that allow you to filter the sample as you are drawing it up into the syringe. The filters add to the cost of the analysis but generally are well worth the investment in protecting the primary separation column. You should always use some form of safe sample injection as part of good laboratory practice protocols.

Once the sample is in the syringe, you are ready to place it in the injector or autosampler vial. For manual injection, place the injector in the load position, push the sample into the injector loop slowly to prevent sample overflow, and then switch rapidly to the inject position, to avoid pressure spiking. Injectors come with a variety of sample loop sizes that can be used for partial loop injections. A good rule of thumb is to inject no more than 75% of the loop capacity or to overfill the loop by 20% when injecting a full loop. After the loop has been washed with six loop volumes, you should return the loop to the load position. I recommend washing the injection needle and syringe with the injection solution three time after injection, to avoid carryover.

5.4   CARTRIDGE COLUMN CLEANUP

Two other techniques that aid column cleanliness are in-line guard columns and off-line cartridge columns to clean samples. Guard columns are short, inexpensive HPLC columns with the same packing support as that of the analytical column. Anything that will pass through a guard column will pass through the main column. Guard columns can be taken off-line, flipped end for end, and washed into a waste container with strong solvent to remove retained organic material and particulate matter. They can be used in place of a particulate filter, but that is not their intended purpose. Compounds that are retained on guard columns will eventually elute down the main column, so they do not protect against organic material that can elute off the main column. People interested in very fast separations who are willing to sacrifice resolution have used guard columns as short analytical columns. Make sure that guard columns for this use have high-resolution particle diameters. Some guard columns will have the same bonded phase bound to large particle diameters and can mess up the chromatography.

Cartridge columns called *sample filtration and extraction* (SFE) *columns* are disposable devices filled with packing material containing the same bonded phase as the HPLC column packing but usually having a much larger particle size. Their initial application was to retain and dispose of material that normally would have to be washed off the column with copious amounts of solvent. They have proven much more useful for preparing a sample "heart-cut," which can then be injected onto the main column to provide a much cleaner sample for analysis.

SFE columns are prepared most commonly from dry packing trapped between two filters in disposable syringe barrels. A octyldecyl silica cartridge first has to be wetted with methanol or acetonitrile that is pushed into a waste container using the syringe plunger. Next, the packing is flushed with starting mobile phase, and then the sample in this phase is pushed through and retained on the packing. The material on the cartridge can now be washed with organic/buffer 1 to elute early running, poorly retained peaks, then with organic/buffer 2 to remove the heart-cut, and finally, with organic/buffer 3 to wash the remaining retained material off the column. Fractions 1 and 3 are checked with an HPLC to make sure that the desired peaks have not been lost, and then fraction 2 is injected onto the LC/MS as the analytical sample. This windowing technique is described in detail in *HPLC: A Practical User's Guide*. The benefit to the separation is less contamination of the main column with strongly retained material that will change the column's running character over a series of injections and require washing out with strong solvent. The chromatography after cartridge column cleanup is usually much simpler, there is less difficulty with bleed from contaminating materials, and total run times are usually reduced. Finally, the cartridge column removes particulates larger than the 30 μm trapping filter pore size of the SFE column.

5.5   ON-COLUMN SAMPLE CONCENTRATION

The more concentrated the sample in the column effluent, the less solvent remains to be removed by the ionizing interface. One of the first justifications for the use of microcolumns was to concentrate the sample while reducing the effluent flow rate to the mass spectrometer. Three primary factors control the concentration of the sample in the effluent output: Increased column efficiency provides narrower chromatography bands; shorter run times with the same resolution prevent band diffusion: and a more concentrated sample at the column head leads to a more concentrated effluent. Column efficiency is controlled by the diameter of the packing material, the regularity of the particle shape and size distribution, and the length of the column. Smaller-diameter spherical packing yields higher efficiencies but also higher backpressures. Efficiency increases with column length, but so do run times, diffusion, and backpressure. Shorter columns with smaller-diameter packing are used to optimize peak concentrations in the effluent. The temperature stability of zirconium-based columns allows elevated temperatures to be used to increase mass transfer into the packing material pore and to shorten run times. Small changes in inlet line and column temperatures can dramatically reduce run times while maintaining peak resolution and shape.

The variable most often ignored is sample concentration at the head of the column. A nonpolar bonded phase concentrates nonpolar components at the column head out of a more polar injection solution. This can be both an advantage and a disadvantage. Concentrating the sample provides sharper bands with less spreading as they move down the column. But if your sample solution is saturated at injection, it can supersaturate and precipitate on the column, although with the limited sample size used for analytical columns, this will generally not lead to column plugging. Start with an equilibrated column and an injection sample that is 50% saturated; then inject, wash with as little starting mobile phase as necessary to remove the early running peaks, and run the sharpest gradient needed to maintain separate target peaks.

5.6   ISOCRATIC AND GRADIENT METHODS DEVELOPMENT

Generally, the separations you will be using for LC/MS have already gone through HPLC methods development. Your primary contribution in adapting them for use in the LC/MS will be to get rid of nonvolatile buffers and ion-pairing mobile-phase components. Occasionally, you may want to replace the stronger mobile phase with one that is more volatile to assist the evaporative interface, but you may have to adjust the stronger solvent concentration to obtain similar run times.

Two types of separations are described in the literature: *isocratic procedures*, in which the mobile phase does not change during the run, and *gradient methods*, in which the mobile phase is changed in a regular and reproducible manner. Isocratic methods are usually preferred for separating simple mixtures of compounds because column reequilibration is not necessary and the effluent reaching the evaporative interface is consistent except for changes in peak components. However, the column may not be able to separate the target compounds in a mixture under isocratic conditions, or the compounds that are retained longest may take too long to elute off the column. Gradient procedures often allow separations that cannot be achieved with a simple mobile phase and allow washing out of late-running peaks, but columns must be reequilibrated to the starting mobile phase before the next sample can be injected. Gradients most run slowly enough to be reproducible and generally will test your pumping system's performance. An inefficient pumping system will give poorly reproducing gradient performance and can be checked by running sample step gradients into a UV detector at 254 nm with one of the solvent reservoirs spiked with acetone as described in Chapter 4. Detection of the acetone will provide a method of visualizing the baseline changes. Bad gradient formation will not show perpendicular slope steps and will show overshooting of the plateaus. This is covered in detail in *HPLC: A Practical User's Guide*.

The resolution of a separation is controlled by three factors: retention (K'), separation (α), and efficiency (N). The resolution equation ties all of these factors together in a single equation:

$$R = \frac{1}{4}\left(\frac{\alpha - 1}{\alpha}\right)\sqrt{N}\,\frac{K'}{1 + K'}$$

where (see Figure 5.1)

$$K' = \frac{V_A - V_0}{V_0}$$

$$\alpha = \frac{V_A - V_0}{V_B - V_0}$$

$$N = 16\left(\frac{V_A}{W_A}\right)^2 = 5.42\left(\frac{V_A}{W_{0.5}}\right)^2$$

Polar components of the sample come off first on these columns, with non-polar compounds being retained the longest. Remember: "Like attracts like." By comparison, polar normal-phase columns retain polar components while releasing nonpolar compounds first. The longer a compound is retained on a column, the better the chance that it will be separated. The retention factor is a measure of the retention of a compound on a column relative to the void volume of the column (Figure 5.1).

Increasing the amount of the stronger, nonpolar solvent used with a octylde-cyl–silica column can decrease retention times by decreasing the mobile-phase polarity. The relative positions of the eluting peaks will remain the same, but compounds will not stick as tightly and will wash out faster. This will help

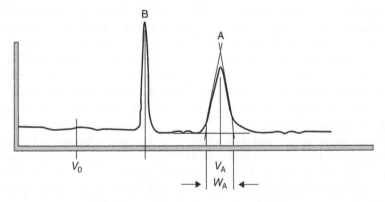

FIGURE 5.1 HPLC separation control factors.

get later runners off faster, but you may lose the resolution of the early-running peaks. The retention factor, K', portion of the resolution equation is a convergent term meaning that after you increase the retention past a certain point, diffusion overcomes the gain you get from making things stick longer on the column.

Selectivity, α, is measured by differences in the retention factors of two adjacent peaks. These selectivity changes cause changes in the relative position of chromatographic peaks. Altering the chemical nature of the column bonded phase, the solvent, or the sample usually changes the selectivity. The easiest change is usually to change the more nonpolar solvent (i.e., methanol, acetonitrile, or tetrahydrofuran) in the aqueous mobile phase. You can also switch to a column with a different organic bonded phase, but this will require buying a second column, downtime to replace the column, and reequilibration time. Although most separations are carried out on C18–silica columns, other columns provide different specificities for different uses. Octyl–silica columns hold nonpolar compounds less tightly and may show peak position switching compared to octyldecyl–silica. Phenyl–silica columns are selective for aromatics and compounds with double bonds, such as fatty acids. Amino–silica columns will separate among monosacchrides, such as glucose, galactose, and fructose, as well as among polycaccharides.

Temperature changes and sample derivatization can also alter selectivity but not in any predictable way useful in methods development. Temperature increases generally will shorten run times while producing separation changes. Temperature as a control variable on silica-based columns always comes with many complications. Bonded-phase hydrolysis and column packing solubility increase with increased temperature, and thermally labile compounds will decompose.

The efficiency factor, N, is the final factor influencing resolution. The efficiency factor measures the sharpness of peaks and the decreasing overlap between adjacent peak bottoms and is measured by dividing the retention volume by the peak width. It is controlled by column packing diameter and shape, packing technique, flow rates, extracolumn tubing volumes, and column length. It is a square-root factor in the resolution equation, so large changes in efficiency do not produce a proportional change in resolution. Columns are often sold on efficiency comparisons, but this can be very deceptive. Efficiency values can change dramatically with the instrument the columns are run on, the methods used for calculation, and the standards selected. Efficiency measurements are valuable for studying changes in column performance over time and definitely should be measured under a standard set of conditions when new columns are being introduced into an accepted method. Following the nature of changes in the separation under these standard conditions over time provides information on the rate of column aging and cleanliness.

Solvent gradients use retention changes to improve separations that cannot be accomplished with a constant solvent. They can be run as a series of step gradient changes to speed elution of later-running peaks. Or they can be run as a continuous solvent change with a final plateau to ensure that all peaks are eluted, followed by a reequilibration step to prepare for the next injection. Timed changes in the gradient slope can be incorporated in the gradient change to improve separation of compressed or expanded areas in the separation. Finally, scouting and washing gradients with other solvents can be incorporated in a multipump gradient system. Generally, only two solvents will be used in an analytical gradient, although you may see a procedure with a constant level of a third solvent added. Developing three- and four-solvent gradients is very complicated, and results are difficult to predict.

5.7   AUTOMATED METHODS DEVELOPMENT

Like everything in modern science, HPLC methods development has been subjected to computer-directed automation. A computer-controlled gradient HPLC system can be equipped with an autosampler and programmed to make a series of injections of the same sample under a variety of gradient conditions. The digitized signal from the system detector is then sent to the controlling computer, and detection is made of baseline deflection maxima, indicating peaks and valleys between peaks. The software then changes running conditions and gradient slopes and continues to inject samples until the best chromatography conditions are determined. The objective is to obtain a maximum number of peaks in a minimum run time. You will have to put an expectation on the range of solvent concentrations to be used, the number of peaks, and the anticipated run time because computer software tends to be monomaniacal in carrying out instructions. Left to itself, the computer will seek an infinite number of peaks in an instantaneous run time.

There are many different software procedures for finding the best run-time conditions. Each equipment manufacturer seems to have adopted a different computerized approach, and the method you choose probably will be locked in by the HPLC system you selected. All of the methods work, but some work more efficiently than others, requiring less operating time and less sample to reach a decision. If you are going to use automated methods development, you need to look at software performance and talk to other experienced users before you buy a system. All software uses the resolution equation and modifications of the retention factor to optimize the final results. It is important to inspect initial results to make sure that the computer has not gotten off on the wrong track. Modifying the solvent boundaries or widening the expected run time can usually pull things back into the real world of chromatographic separations. Methods development can be shortened dramatically if you can put an actual run condition or a best guess set into the software as a starting point. I sold and use many different types of optimization software and was most impressed with the simplex optimization or "drunkard's walk" approach. Many other software systems seem to be dedicated to the "infinite monkey" theory of development and run all possible points within the given boundaries, which can be very time consuming and expensive. These automated methods are best run off-line on a secondary detector and not taken to the ionization interface and the mass spectrometer until they have been worked out sufficiently.

6

LC/MS INTERFACES

Once compounds have been separated into distinct chromatographic peaks, they need to be moved into the mass spectrometer for analysis. This includes everything coming out of the column: separated compounds, elution solvent, and volatile reagents. Solvent and contaminates must be evaporated, and the target compounds in the peaks need to be ionized in an interface so that they can be drawn into the mass spectrometer analyzer.

6.1 SOLVENT REMOVAL AND IONIZATION

The flow rates produced by an HPLC will overwhelm a high-vacuum mass spectrometer pinhole interface. The mass spectrometer analyzer is operated at 10^{-5} to 10^{-7} torr vacuum pressure to prevent collision with air molecules of the molecular ions produced in the interface, which could cause them to fragment before being detected. Sucking in all the output from the HPLC column would overload the capacity of the vacuum pumps completely swamp out the output signal coming from the analyzers. The interface must be capable of removing most of the volatile solvent and its additives without removing the target compounds of interest. This fine balance is accomplished by using a combination of heaters, reduced pressure, and gas nebulizers to volatilize and entrain the unwanted eluant solvent components.

Splitting the eluant stream in the interface can reduce the total input to the mass spectrometer if the sample concentration is high enough to allow

LC/MS: A Practical User's Guide, by Marvin C. McMaster
Copyright © 2005 John Wiley & Sons, Inc.

the signal to be detected by the analyzer. A solvent diversion valve placed in the flow path either before the sample enters the interface or after the sample enters the ionizer module will produce the required split. A micro- or nanoflow HPLC system designed to use a very small-diameter HPLC column and reduced flow rates also decreases effluent feed to the interface. Increased signal and a reduction in the amount of sample needed for detection is the main claim to fame for these. The electrospray interface tends to be very flow dependent and gives the best signal with very slow effluent and benefits most from nanoflow systems.

The second main function of the interface is to ionize the target compounds, since many of the components in the HPLC eluant are uncharged. The mass spectrometer is designed to move charged molecules of a selected mass-to-charge ratio from its analyzer entrance to the ion detector. Ionization of noncharged molecules is accomplished by spraying the solvent droplets either from an electrically charged capillary or across a coronal discharge needle during the final stages of evaporation.

6.2   ATMOSPHERIC-PRESSURE INTERFACES

Modern LC/MS systems come equipped with nonevacuated interfaces in which solvent is being removed as the samples are ionized for introduction into the high-vacuum environment of the mass spectrometer analyzer. The two types of in-flow interfaces in common use in modern LC/MS systems are the electrospray (ES) and atmospheric-pressure chemical ionization (APCI) or ion spray (IS) interfaces. Most commercial systems are equipped with both interfaces, or a user can switch between interfaces, which are available as an option when a system is purchased. They also allow switching the ionization polarity of the interface either to negative ionization mode, to provide deprotonated ions, or to positive ionization mode, to give molecular ions or solvent adducts. Both of these atmospheric-pressure interfaces are rugged and designed to prevent contamination of the mass spectrometer. Either ES or IS interfaces can serve as the front end for tandem mass spectrometers that can separate elution ions in the first mass spectrometer and then fragment the separated ions to form daughter ions that can be analyzed in the second MS system for identification and structural studies.

Another interface commonly used for connecting HPLC to a mass spectrometer is not a true in-line interface. It is a robotically controlled spotter plate system for collecting samples from the HPLC to be injected into a MALDI (maser-assisted laser desorption and ionization) time-of-flight mass spectrometer. The effluent sample dropped in the plate well is mixed with the ionization matrix already present, solvent and volatile reagents are evaporated, and the plate is then placed into the injector target. There it is blasted with a pulsed laser to volatilize and throw the ionized sample into the atmosphere of the interface, where it can be drawn into the mass spectrometer.

6.3 ELECTROSPRAY INTERFACE

The electrospray (ES) interface (Figure 6.1) is recommended for use with highly polar and ionized materials. It is a very soft ionization technique that results in little fragmentation. It is concentration independent, thus lending itself to microflow miniaturized HPLC systems.

In the electrospray interface, ionization is accomplished by passing the HPLC eluant down a heated metal capillary tube along which a 3- to 4-kV electric charge differential is applied. The evaporating liquid is sprayed out of the end of the tube as charged droplets decreasing rapidly in size. In some models the charged tube is surrounded by another tube, containing a nebulizing inert gas, to aid in final evaporation.

With most proteins, an ES interface produces multiply charged ions. The normal molecular-weight ranges of benchtop systems allow separation of the ions produced because the analyzer separates on the basis of mass divided by charge number (m/z). The spectrum produced from the protein is made up of an envelope of peaks with various m/z ratios. Deconvolution software can analyze this peak envelope to determine the molecular weight of the uncharged protein. The nanospray interface is an electrospray interface optimized for use with microflow HPLC systems.

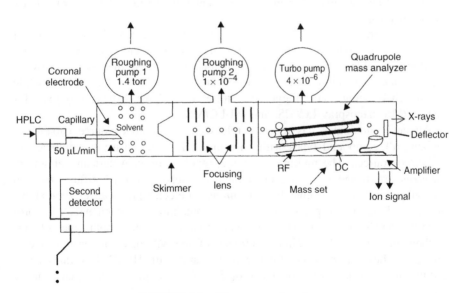

FIGURE 6.1 Electrospray interface.

6.4   ION SPRAY INTERFACE

The ion spray (IS) interface, which is used with less polar effluents, is the
workhorse for standard HPLC systems since it can take flow rates up to 2 mL/min.
It is most commonly used to produce intact molecular ions for molecular-weight
determinations, but it can be set up with an ion repeller to cause fragmentation,
which can provide preliminary compound and structural information.

The original IS interface used an impactor plate offset from the pinhole
entrance to the mass spectrometer. Solvent that was not volatilized hit the plate,
collected, and ran off to a drain leading to a bucket that collected unused effluent.
Use of a solvent diversion valve allows effluent not used in the MS interface to
be sent to a fraction collector or to a secondary traditional HPLC detector for
nearly simultaneous analysis to provide additional information about the separated
peak components.

Modern LC/MS interfaces continue to use a convoluted path or curtain gas to
ensure that only charged sample enters the mass spectrometer entrance. Unused
solvent is drawn off with a diversion valve to a secondary detector for identity
confirmation and collection.

The IS interface (Figure 6.2) also uses a nebulizing inert gas to entrain and
break up the eluant stream into small droplets that are sprayed across a coronal
discharge needle operated at about 25 kV to ionize the shrinking droplets. The
impactor plate is equipped with a charge opposite to that applied to the coronal
needle, to draw the charged ion to the mass spectrometer entrance. Again, the
nebulizer capillary may be heated to aid evaporation, or an oppositely charged
plate may draw the charged droplets into a heated tube before they enter the mass
spectrometer inlet.

The problem with the ESI, nanospray, and ISI interfaces is that they each fail
to ionize some types of compounds. To solve this problem, users must acquire
interfaces capable of ionizing their compounds of interest, then switch between
the interfaces as required. This is time consuming since the system is down while
the switch is being made. Large laboratories often have dedicated systems for
ESI-LC/MS, nanospray-LC/MS, and ISI-LC/MS.

AligentTechnologies has recently released a multimode source capable of
ionizing all compound types with only minor loss in sensitivity compared to
dedicated sources. The multimode source, reported in an American Laboratory
magazine article and on www.chem.agilent.com, leads the HPLC effluent through
a nebulizer into a single chamber containing an electrospray ionizer zone above
an infrared thermal vaporization chamber containing a corona needle to produce the APCI ionization. The combined stream containing both types of ions
is then directed to the capillary entrance of the MS detector through a curtain gas. They report successful use of the unit with HPLC flow rates up to
2.0 ml/min. and detection sensitivity of 5 pg of Reserpine with signal to noise
ratio of 25.

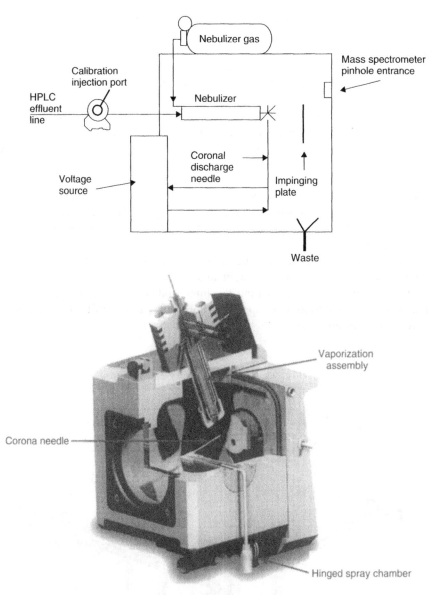

FIGURE 6.2 Ion spray interface. (Courtesy of Varian.)

6.5 SECONDARY DETECTORS

Solvent diverted from the interface can be sent to a secondary detector to provide further information about a sample, or it can be sent to a fraction collector for collection. Mass spectral analysis is a destructive technique using the current crop of detectors, but very little sample is required to provide the MS signal. HPLC is both an analytic and a preparative technique, and most HPLC detectors can provide information without degrading the sample.

By far the most versatile and commonly used HPLC detectors are the ultraviolet (UV) detectors. They operate on the principle that most organic compounds absorb ultraviolet light at some wavelength. Detectors can be set at wavelengths specific for a particular compound's absorption, set at a low wavelength to detect almost any compound, or run in a photodiode detection mode to scan a broad range of UV wavelengths, providing an array of information from an HPLC stream nearly as complex as that from the mass spectrometer. The UV signal is not dependent only on concentration but also on the absorption coefficient of the chromophores present in the effluent samples. They can provide additional information on the presence and identity of the target compounds and any impurities.

A secondary fluorescence (FL) detector can provide specific information about compounds that can absorb UV light and promote it to a higher wavelength for fluorescent emission. By tuning the input and detection wavelengths, these detectors can be set to look for specific compounds in the HPLC stream, usually at very high sensitivity.

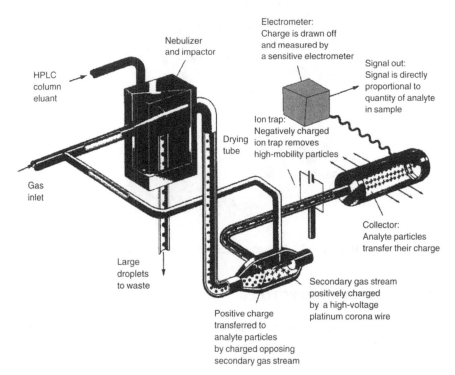

FIGURE 6.3   Corona charged aerosol detector. (Courtesy of ESA.)

A refractive index (RI) detector is a concentration-dependent detector that looks for compounds with a refractive index that differs from the mobile phase. It can see almost any compound present in the flow cell but is limited by low sensitivity and use to isocratic solvent flows.

Other HPLC detectors, such as conductivity and electrochemical detectors, are generally not used as secondary detectors because they tend to destroy a sample or require high ion concentrations.

A detector that has recently appeared on the HPLC market may provide a nice complement to mass spectrometer detection. It is offered as a high-sensitivity universal detector for gradient work using volatile buffers. The corona charged aerosol detector uses something like a ion spray nebulizer to evaporate solvents and buffer, then places a charge on a multiatom droplet formed by passing it over a low-voltage charged needle. The charge on the droplet is then measured with an electrometer. It is advertised as providing good detection of carbohydrates, phosholipids, steroids, and peptides that are difficult to measure with a UV detector (Figure 6.3).

7

LC/MS OVERVIEW

The basic components of the LC/MS system are an HPLC pumping system, a separation column, an ionization interface, an MS:vacuum system, focusing lens, an analyzer, an ion detector, and a data/control system (Figures 7.1–7.4).

Vacuum pumping systems for a mass spectrometer provide the high vacuums ($<10^{-5}$ to 10^{-7} torr) critical to the operation of the spectrometer by preventing collision of ions with air. Ionized compounds cannot exist long enough to reach the detector if they suffer collisions with air molecules in the analyzer. Vacuum is achieved in two stages: first a roughing pump to drive the vacuum to 10^{-4} torr and then a finishing pump to move on down to 10^{-5} to 10^{-7} torr. Pumping is done in series, with the output from the finishing pump evacuating the analyzer feeding the roughing pump and then out to the atmosphere. Dual finish pumps, one evacuating the mass spectrometer source and the second providing vacuum only for the analyzer, may be provided for high-volume or high-flow-rate applications.

Four basic types of LC/MS systems are in common use in analytical laboratories. The first and oldest type is the *quadrupole* or its cousin, the *octapole, analyzer*, which use a bundle of oppositely charged rods swept with radio frequencies to separate and focus ions from the injector on the detector. The second system type uses either a three-dimensional spherical or a quadrupole *ion trap* to contain and then release ions as the radio frequency changes from the trap to the ion detector for mass determination. The third type ejects ions in a burst mode into a *time-of-flight tube*, which separates ions by their mass-to-charge ratios and focuses them on its detector. The fourth system type has a trapping volume that is bombarded with a full-frequency radio signal that momentarily promotes ions

LC/MS: A Practical User's Guide, by Marvin C. McMaster
Copyright © 2005 John Wiley & Sons, Inc.

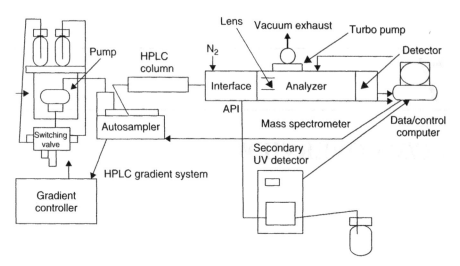

FIGURE 7.1 LC/MS system diagram.

to a higher-energy state and then collapses, producing a modified all-frequency signal that is detected and analyzed using *Fourier transform* software to give an intensity versus m/z signal. Combinations of one or all of these mass spectrometer system types are used to create an *MS/MS family* of hybrid systems that combines multiple mass spectrometer analyzer modules with collision and ion-trapping cells to separate, fragment, and detect by their molecular-weight differences daughter ions of the original molecular ion. The ability of MS/MS systems to analyze the fragmentation products of molecular ions allows structure analysis studies for confirming identification of the materials separated by the HPLC beyond simply providing molecular-weight information.

The primary use of MS modules in most laboratories at present is the determination of the molecular weights of compounds separated by the HPLC portion. A second use is to provide conclusive identification of the structure of the separated material by structural studies or by comparison to fragmentation libraries. A third use is to prove the presence of or to identify coeluting impurities. Finally, accurate-mass mass spectrometer modules are used to confirm elemental analysis by isotope ratio studies. This application requires the use of specialized high-resolution analyzer systems and is beyond the scope of this book.

7.1 HPLC AND THE IONIZATION SOURCE

The interface between the HPLC and the mass spectrometer is critical for system performance, as covered in some detail in Chapter 6. All interfaces are designed to transfer ionized samples from the HPLC effluent into the source without mixing separated bands of compounds. In most LC/MS systems the interface configuration can be changed to produce ions with either positive or negative charges from either an ion spray or an electrospray input. These ion molecules are pulled into the analyzer of the mass spectrometer across a pinhole orifice by the pressure differential between the atmospheric interface and the high-vacuum analyzer region. Interfaces are designed to ensure that nonionized sample is diverted away from this entrance pinhole either by convoluted impinging pathways or by using a curtain gas. Occasionally, interfaces are equipped with a repeller plate charged with the same charge as the ion molecule, to force them toward the pinhole entrance. These should be used with care because they can add energy and momentum to the already charged ion, leading to fragmentation that can complicate the ion current spectrum interpretation, making molecular-weight determination more difficult.

7.2 VACUUM PUMPS

The first requirement for mass spectrometer operation is a high-vacuum environment. The ionized analyte ions formed in the interface between the HPLC and the mass spectrometer cannot survive collision with air molecules in the analyzer. A rough vacuum is established first with a laboratory rotary-vane oil pump, and then a turbomolecular pump connected in series provides the final operating vacuum. Turbo pumps have a series of vanes mounted on a shaft that rotates at very high speeds. The vanes entrain air from the vacuum chamber and exhaust it out through the roughing pump. A turbo pump is effectively a jet engine for the mass spectrometer.

The turbo pumps in many desktop systems are only the size of your doubled fist, but they can bring the analyzer pressure down to 10^{-6} torr on a good day. They do not have the pumping capacities of larger turbo pumps that can move 150 to 2500 L/s. They are used on systems that do not have high source pressures. The biggest advantage of a turbo pump is that it contains no oil to contaminate the analyzer. Its biggest drawback is mechanical failure, although since the introduction of turbo pumps, their performance life has constantly been improving. Always work with a manufacturer that has a good trade-in program. You definitely do not want to get involved with maintenance or rebuilding of one of these pumps.

Oil diffusion pumps are found on older and larger floor-standing systems. These pumps heat oil that rises and entrains gases, then encounters a coiled chiller where the oil is cooled, releasing the entrained gas to the roughing pump. The oil is returned to the reservoir down the wall jacket for recycling. Cooling systems are massive, and a diffusion pump needs to be protected from dumping oil into the analyzer if power is suddenly lost.

7.3 ANALYZER AND ION DETECTOR DESIGNS

Once a sample is ionized, it must be focused and drawn into the analyzer portion of the mass spectrometer, with charged drawout and a focusing lens electrically

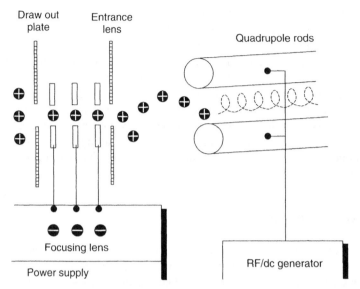

FIGURE 7.2   Focus lens.

charged with the same charge type as the ions to concentrate them into a beam (Figure 7.2).

The direct-current (dc) charged surfaces of the analyzer are then swept with a changing radio-frequency (RF) signal that selects for different mass ions for each frequency, allowing them to follow a stable path to the detector. The stable ions at each frequency are then expelled to collide with the ion detector to be counted by striking the surface of the detector module, generating a signal that is amplified and sent to the computer.

The quadrupole analyzer has four cylindrical rods clamped in a tubular arrangement by ceramic collars. Opposing rods have the same dc charge applied to them, while adjacent rods have the opposite charge applied (Figure 7.3).

Ions are focused into the tunnel formed by the four rods and follow a corkscrew flight pattern down the rods as they are swept forward by the changing RF signal. Ions masses that are not selected by the combined dc/RF signal follow unstable paths, collide with the inner walls of the analyzer rods, and are lost. The ion detector most commonly used with quadrupole and ion trap systems uses an impact/cascade detection mode (Figure 7.4).

When a charged particle strikes a detector's membrane surface it causes an electron to be released from the other side. These electrons then strike the coated walls of the detector, releasing multiple electrons on each impact. This cascade of electrons amplifies the signal of a single contact for transfer to the data system. In the data system, the data received are converted to a chromatogram of signal strength versus elapsed time as a total ion chromatogram (TIC).

The mass spectrometer also knows the m/z value of each time point from the RF sweep signal sent to the analyzer. These data are combined for a number

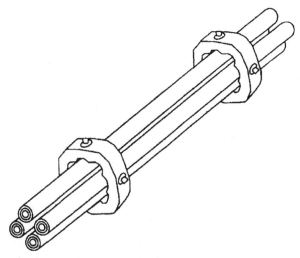

FIGURE 7.3   Quadrupole rods.

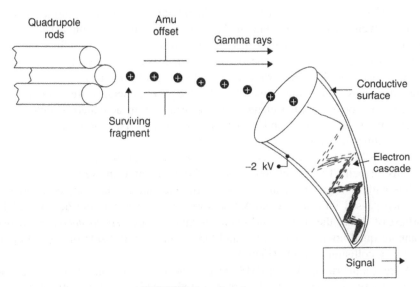

FIGURE 7.4   Ion detector.

of scans and provide the molecular weight of the molecular ion. The *m/z* value determined for each peak may be displayed above the peak, but this depends on the software of the system being used. In other cases, a table is generated displaying *m/z* versus retention times versus total ion signal strength.

In ion trap systems all sample ions injected are held in a stable circular path in the analyzer by a constant-maintenance RF signal between a sandwich of dc charged plates of opposite polarity. The paths that the ions follow are said to resemble the stitching on a baseball. Changes in the radio-frequency signals as the dc/RF signal is swept to higher or lower frequencies cause each mass ion's path in turn to become unstable, releasing it to the ion detector below the center of the bottom analyzed plate. Again the number of impacts at each frequency are measured, amplified, and sent to the computer. One advantage offered by an ion trap analyzer is the ability to trap and hold specific ions between the analyzer plates and then to induce collision fragmentation with gas introduced into the trapping chamber, followed by dc/RF frequency changes to release fragmentation ions to the ion detector. This allows a single analyzer to act as an inexpensive MS/MS system for fragmentation structure studies.

Time-of-flight analyzers are almost in a class by themselves. More conventional time-of-flight systems use a standard atmospheric pressure ionization interface and a trapping cell to provide bursts of ions to the flight tube for separation and detection by an ion detector. The molecular ions striking the detector surface induce a cascade of electrons within the detector body that amplify the single fragment signal, sending a stronger signal to the detector electronics. This signal can be amplified further with an electron multiplier tube, to provide a strong enough signal for data system processing.

In a MALDI/time-of-flight system, effluent sample is mixed with a chromaphore, solvent is dried off, and the mixture is bombarded with a laser beam to form ions. Ions are released in a burst into a flight tube in which ions are separated by the flight times needed for each mass ion to reach the detector. Again impacts on the detector are amplified and sent to the data system as signal strength versus flight time. Time-of-flight analyzers generally use a diode-array impact detector, with each element in the array being activated to detect at a specific time. This signal can be converted to signal versus mass by comparison to flight times of calibration standards.

The MS/MS analyzer usually combines two mass spectrometer analyzers with a collision cell. The target ion selected from the first analyzer is allowed to collide with inert gas molecules to induce fragmentation. The fragmentation ions are then passed into the second MS analyzer for separation and detection. The earliest of these, the *triple-quadrupole* or *tandem mass spectrometer*, used two scanning quadrupole modules, Q1 and Q3, units on either side of a quadrupole used as a collision cell, Q2 (Figure 7.5).

A triple-quadrupole LC/MS/MS system can be run in one of four modes to run a variety of experiments (Figure 7.6). Other MS/MS combinations can run in scan/scan mode, but may have trouble running all four of these experiments. For instance, a three-dimensional ion trap has problems running a neutral loss scanning or MRM (multiple-reaction monitoring) experiment unless it is part of a hybrid system.

In the *daughter mode* (1), Q1 is scanned and all ions are sent to the fragmentation chamber, Q3 is then parked at a specific m/z frequency to look for a specific fragmentation ion common to related Q1 ions. In the most common *parent mode* (2), Q1 is parked at a specific frequency to select only one ion to send to the collision chamber, and Q3 is scanned for fragmentation information

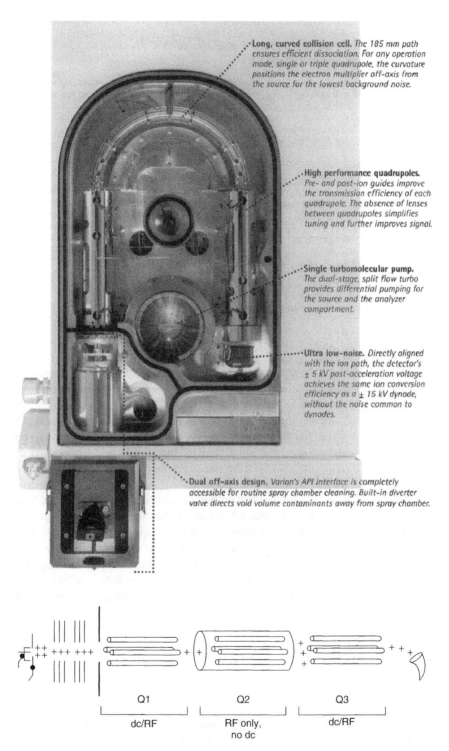

Long, curved collision cell. *The 185 mm path ensures efficient dissociation. For any operation mode, single or triple quadrupole, the curvature positions the electron multiplier off-axis from the source for the lowest background noise.*

High performance quadrupoles. *Pre- and post-ion guides improve the transmission efficiency of each quadrupole. The absence of lenses between quadrupoles simplifies tuning and further improves signal.*

Single turbomolecular pump. *The dual-stage, split flow turbo provides differential pumping for the source and the analyzer compartment.*

Ultra low-noise. *Directly aligned with the ion path, the detector's ± 5 kV post-acceleration voltage achieves the same ion conversion efficiency as a ± 15 kV dynode, without the noise common to dynodes.*

Dual off-axis design. *Varian's API interface is completely accessible for routine spray chamber cleaning. Built-in diverter valve directs void volume contaminants away from spray chamber.*

Q1   Q2   Q3

dc/RF   RF only, no dc   dc/RF

FIGURE 7.5 Triple-quadrupole MS/MS system. (Courtesy of Varian.)

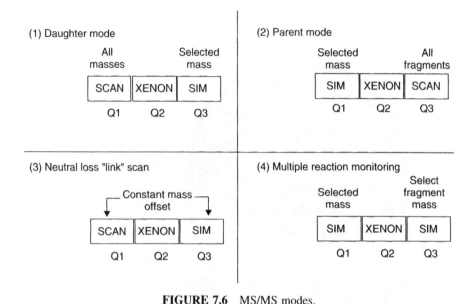

FIGURE 7.6   MS/MS modes.

that can be used to identify the structure of the ion from Q1. In the *neutral loss* or *linked scan mode*, 3, both Q1 and Q3 are scanned with a specific frequency offset. Ions that are detected must have lost a common uncharged molecule that can provide information as to their fragmentation type. The last experiment (4), used to investigate trace compounds in complex mixtures, is less common. Q1 and Q3 are set at single-ion frequencies specific for the impurity to be analyzed and one of its daughter fragments. Since most of the signals being generated are ignored—like focusing only on specific trees in a forest—very specific signals can be generated for traces of material present.

MS/MS systems are not limited to combining analyzers of only a single type. Hybrid systems can be made to take advantage of the strengths of different ana-lyzer types. One common analyzer combination uses a quadrupole, Q1, as a front end to feed a time-of-flight, Q3, analyzer. One of the newest combinations is the Qtrap system, combining a quadrupole module, Q1, with a linear ion trap, Q2. The ion trap is based on a set of quadrupole rods and can be run either as a triple-quad system or used as a quad trap in trapping combinations that a triple quad cannot achieve.

7.4   DATA AND CONTROL SYSTEMS

The data system records the signal strength of all ions being released to an ion detector at a given time. A data system stores all the spectral information from the chromatographic run in a three-dimensional block of data. An *ion current chromatogram* or *total ion chromatogram* (TIC; Figure 7.7a) is a plot of the sum

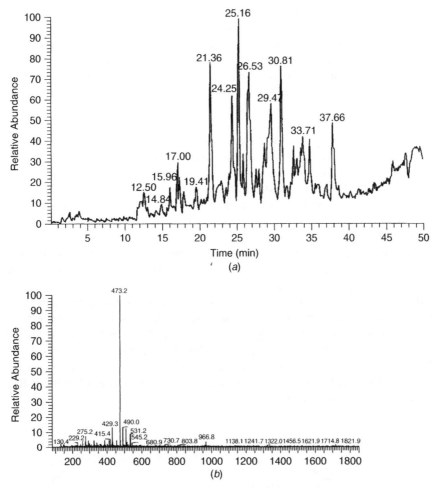

FIGURE 7.7 (*a*) Total ion chromatogram and (*b*) *m/z* spectrum. (From Tiller et al., 1997.)

of the signal strengths of all ions present versus time. A *spectrum* (Figure 7.7*b*), is a plot of signal strength versus mass/charge (*m/z*) at a given time. For an IS interface system, this will be a single mass for the molecular ion. In an ion trap system we can read the molecular ion, flood the chamber with a heavy gas, fragment the molecular ion, and then read the ion fragment masses. If we are displaying fragmentation data, the spectrum will be a bar chart of all the fragment masses and their signal strengths. The frequency of the dc/RF signal is calibrated with standards of known mass and charge to provide the *m/z* value.

It is obviously advantageous to acquire as much information as we can about a fragmentation sample the first time we run a new chromatogram on unknown material. We can set the mass spectrometer in scan mode to acquire data from a range of dc/RF frequencies wide enough to cover the expected range of m/z values while excluding low mass values from air and solvent. We need to set a signal-sampling rate and allow for the length of time to make the chromatographic run, and all these data have to fit in the data storage space available. The greater the number of sample points we can average at a given m/z frequency and time, the more confidence we will have in the data point. If we want to scan a wider m/z range, we may have to use a lower sampling rate and reduce the accuracy of the signal.

The signal coming from the mass spectrometer's detector is a continuous voltage changing with time, an analog signal. Data processing in a computer works with discrete bits of digital data. We must convert the analog signal from the detector into a digital data stream to use it in our computer system. The signal conversion process involves measuring the vertical displacement of a series of time slices of known duration using an A/D conversion microprocessor card (Figure 7.8).

The value of the time duration and the changing values for the signal intensity are stored as the data set. A useful analogy is the creation of a motion picture. The moving image projected on a screen is created by viewing a series of changing still photo frames equivalent to the time-slice data taken from the analog mass detector signal. The still photos represent digital data combined to produce the analog motion picture. Data from the mass controller signal selecting the m/z range's change with time is combined with the detector voltage versus time database to provide the information the computer needs to produce a total ion chromatogram, and at any given time, the mass spectra made up of the voltage signal versus the m/z value.

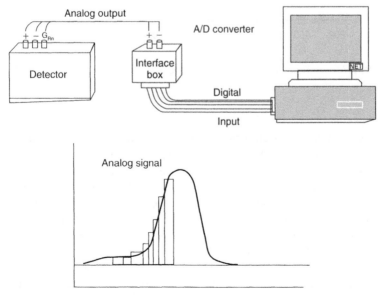

FIGURE 7.8   A/D data conversion.

7.5   PEAK DETECTION, ID, AND QUANTITATION

When we are operating an LC/MS system, we usually set the data system so that we can monitor the total ion current as a chromatograph on a display screen as it is being generated. Each point we see on a chromatogram is usually the summation of eight or more dc/RF scans. We may want to extract and display a spectrum as the peaks come off and we see a peak of interest, to check its molecular weight or its fragmentation pattern if we are running an MS/MS experiment. Qualitative information can be derived from the spectrum as we analyze the sample. Fragmentation patterns of interest can be sent to the data system for comparison with libraries of known fragmentation patterns to provide definitive compound identification beyond the molecular weight. Quantitative information can be obtained for HPLC chromatographic peaks from either a total ion chromatogram or by using the data system to compare the information to a calibration curve of known amounts of the material of interest versus peak heights or peak areas.

In obtaining data for a known sample, increased signal sensitivity can be obtained by using the single-ion monitoring (SIM) mode by selecting only a single m/z frequency and acquiring data only at that point to produce a single-ion chromatogram. The sampling rate no longer has to be spread over the entire m/z spectrum, and the number of measurements can be increased for this single-ion mass.

The combination of fast analyzer scanning, fast detector recovery, and high-capacity data systems allows acquisition of about 25,000 data points per second. This means that a mass spectrum run in scan mode from 35 to 550 m/z can average 8 to 10 scans in 1 second. Run in single-ion mode the same mass spectrometer could analyze 10 m/z mass regions in a step scan and gain a tremendous gain in sensitivity by averaging a much higher number of points at each of the 10 points.

8

MASS ANALYZERS

The heart of a mass spectrometer is the mass analyzer, which separates ions by their mass divided by their charge, their m/z ratio. These can be molecular ions with positive charges, adducts made from a combination of molecular ions and solvent or mobile-phase components, fragmentation ions from a collision chamber, or negatively charged ions produced when the polarity is switched in the ionization chamber and in the draw-out and focusing lens. The purpose of the analyzer is to hold ions, select specific mass ions as the radio frequency is scanned, and move the selected ion into the ion detector for counting.

8.1 QUADRUPOLE ANALYZER

The quadrupole mass spectrometer was the first type of analyzer used in desktop LC/MS systems and is still by far the least expensive and most common analyzer used in LC/MS systems. It will be the used to illustrate analyzer and detector operation. The first step is moving charged ions from the ionizing interface into the higher-vacuum area of the analyzer through a pinhole entrance. Just past the entrance are a series of electrical repeller, draw-out, and focusing lenses (Figure 8.1). The exact configuration will vary from manufacturer to manufacturer. Their purpose is to get the ions entering the high-vacuum entrance of the mass spectrometer into a condensed stream before passing through the analyzer rods.

LC/MS: A Practical User's Guide, by Marvin C. McMaster
Copyright © 2005 John Wiley & Sons, Inc.

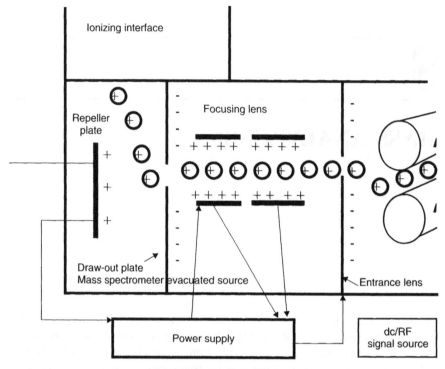

FIGURE 8.1 Focusing lens.

A draw-out lens with opposite charge works with movement from atmospheric pressure to a high-vacuum environment to pull ions into the entrance chamber. A repeller at the back of the entrance chamber has the same charge state as that of the ions and forces them into the focusing lens. Variable voltage charges on the focusing lens with the same polarity as the molecular ions squeeze the ion beam into an intense stream as it enters the quadrupole analyzer.

The quadrupole mass analyzer is the heart of the mass spectrometer. It consists of four cylindrical quartz rods clamped in a pair of ceramic collars. The exact hyperbolic spacing between diagonally opposed rods is critical for mass spectrometer operation. Rods should not be removed from the ceramic collars except by a service organization.

Both a dc and an oscillating RF signal are applied across the rods, with adjacent rods having opposite charges (Figure 8.2). The ion stream entering the quadrupole is forced into a corkscrew, three-dimensional sine wave by the combined effect of the dc applied to the rods and the RF signal sweeping down the quadrupole rods, generating the electromagnetic field of the analyzer. This combined dc/RF field applied to the rods is swept toward higher (or lower) field strength by the computer-controlled dc/RF generator, upsetting the standing sine wave for all but

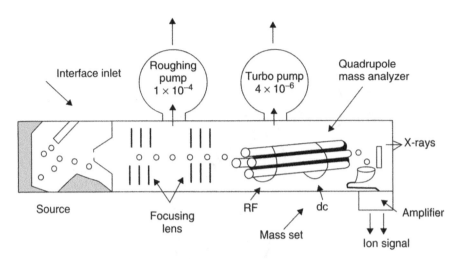

FIGURE 8.2 Quadrupole analyzer.

a single fragment mass at a given frequency. This single mass follows a stable path down the length of the analyzer and is deflected onto the surface of the detector by the amu offset lens. Any ion fragments not passed at a given dc/RF frequency follow unstable decaying paths and end up colliding with the walls of the quadrupole rods, picking up a charge-neutralizing electron. As the dc/RF signal is swept up to higher frequencies, larger masses are led in sequence to strike the detector faceplate.

The fragments that pass the analyzer strike the exterior surface of the detector horn (Figure 8.3) after being deflected away from the linear path out of the analyzer by a lens called the *amu offset*. In GC/MS, gamma particles produced in the ionization source filament are not deflected but would cause false signals if allowed to strike the detector surface directly in the case where the ions were not deflected by the offset lens.

When a molecular ion strikes the face of the detector, it causes electrons to recoil off the coated inner face of the detector horn. These electrons strike the inner side of the horn, producing a shower of secondary electrons. This growing cascade of electrons bouncing down the tapered sides of a detector horn eventually reaches the output terminus, producing a greatly amplified signal, which is then led to an A/D converter providing the signal to the computer system. This signal is scanned at a rate exceeding 35,000 points per second and combined with the dc/RF signal to produce the spectrum and the ion current signal of the mass spectrometer. Plotting this against the run time from the chromatograph yields the total ion chromatogram. Inspection of the spectrum at any given time will provide information on the molecular weights of any compound or fragmentation ion present in the separation at that time.

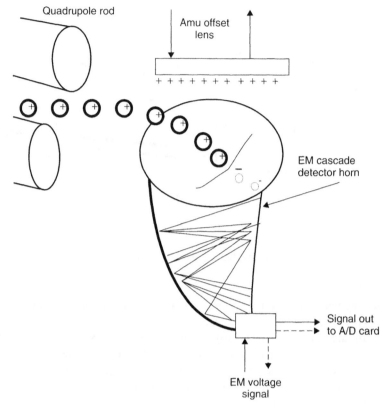

FIGURE 8.3 Ion detector.

8.2 ION TRAP ANALYZER

Ion trap mass spectrometers (ITDs, ITMSs, and linear ion trap systems) are finding growing acceptance in analytical laboratories. Laboratories that use them
claim that they are 10 to 100 times more sensitive than a quadrupole. They can
easily be switched between APCI, CI, and EI modes, they require less maintenance, and they have the potential to be used for MS/MS studies.

The desktop ion trap detector (ITD) and the floor-standing ion trap mass spectrometer (ITMS) vary in size and added function more than in theory of operation.
The ITMS is designed as a research instrument with both analytical MS and
MS/MS operation in mind. The ITD is a dedicated, compact unit with a smaller
trap and pumping system, designed for production GC/MS or LC/MS operation.

Molecules introduced into an ITD ion trap from an atmospheric ionization
source are processed totally within the body of the ion trap. Ionized molecules
enter the trap from a pinhole orifice on the side of the ring electrode and are
held in a three-dimensional spherical-segment stable orbit between the electrodes
with a trapping voltage from the dc/RF source. They are then eluted in increasing

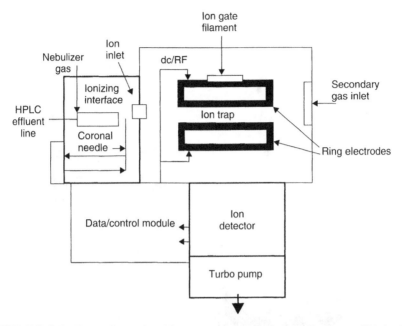

FIGURE 8.4 Three-dimensional ion trap detector analyzer. (Courtesy of Varian.)

mass (m/z) by increasing the dc/RF voltage on the ring electrode. This pushes each fragment ion into an unstable orbit, causing it to escape through one of the seven holes in the exit electrode and into the dynode electron multiplier detector, which sends a signal to the data system (Figure 8.4). The detector horn lies immediately below the exit electrode. The turbo pump is mounted directly below and attached to the ion trap body by a Vitron O-ring gasket, both of which are enclosed in a heated vacuum manifold. Also on the manifold are the attachments for the cal gas valve, an entrance line for chemical ionization gas, and an exhaust port for the rotary vane mechanical pump. The ion trap arrangement allows for the addition of collision gas, such as carbon dioxide, methane, or ammonia. Collision of a specific ionized molecular ion trapped between the ring electrodes with this gas can lead to fragmentation to aid in identification of the target ion's molecular structure.

Only a limited volume of sample can enter the ion trap without overloading and causing performance degradation. Narrow-bore capillary column with flows of 1 mL/min can be directly interfaced, or a splitter column can be used to divert part of the LC stream to a secondary detector. Once the sample is in the trap, if not already ionized in an atmospheric-pressure interface, it can be ionized with 70-V electrons from the ion gate in the entrance electrode at the top of the trap (Figure 8.5).

A heated filament electrode positioned above the trapping ring electrode plates can furnish thermionic electrons for further ionization when inducing fragmentation. Between the filament and an unused spare filament is a repeller plate that

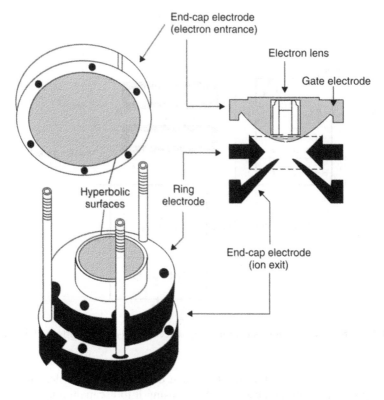

FIGURE 8.5 Ion trap electrodes.

drives the electrons toward the ion trap containment space (Figure 8.6). At the base of the ionization electrode is a variably charged electron gate. When the gate has a high negative charge, electrons stay in the electrode; when the gate goes positive, electrons are forced into the storage space and ionize molecules of the sample.

The ring electrode around the containment space has a constant RF signal of variable voltage. A storage voltage of 125 V ac is applied to trap all ions with mass equal to or greater than 20 amu. At this voltage the ions formed are throw into spherical-segment hyperbolic orbits that are described as resembling the stitching on a baseball (Figure 8.7).

Approximately 50% of all ions formed are thought to be trapped and eventually reach the detector. This compares to the single ion at a given time that reaches the detector in a quadrupole; most ions end up colliding with the quadrupole rods and are never analyzed. This increased ion yield explains the increased sensitivity of the ion trap detector. Some increase in ion trap analyzer stability comes from the lack of sample accumulation on the electrodes, although this will vary from sample to sample. Makeup helium gas is usually added from the collision gas lines and serves an important role in stabilizing the ions in their

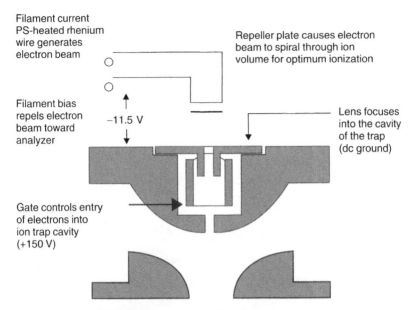

FIGURE 8.6   Ion trap filament and ion gate.

FIGURE 8.7   Stable ion schematic.

orbits. Frequent collisions between the small, fast-moving gas molecules and the charged ions dampen their movement, causing them to collapse toward the center of the trap. Ion orbital stability is also improved by applying axial modulation. This is voltage applied between the ionization electrode and the exit electrode at a frequency about one-half that of the ring electrode voltage. It has the effect of moving ions away from the center of the trap, where the voltage is zero. This

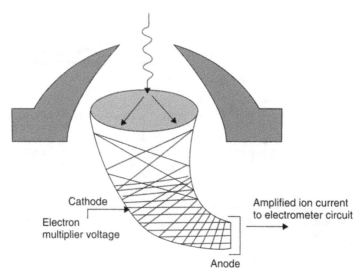

FIGURE 8.8 Exit electrode and detector.

aids in ion ejection from the trap and dramatically sharpens the mass resolution at the detector.

Analysis is performed by gradually increasing the ring electrode's scanning voltage. This upsets the orbits of ions with increasing masses, causing them to escape through the exit electrode's holes and strike the dynode's surface. Scanning is done in four segments over the full scanning range. This allows for mass peak height manipulation and tune modification. With this tool, the tune can be adjusted to meet specific peak ratio requirements. Ion trap fragment scans using four-segment scanning can be used in searching existing database libraries for compound identification.

The ion trap ion detector is the dynode electron multiplier that we have previously seen used in quadrupole systems (Figure 8.8). Positive ions striking its lead oxide glass cathode surface release electrons from its inner surface. These bounce down the inner walls, releasing a cascade of electrons on each contact. As many as 100,000 from a single positive contact will reach the anode cup and send a signal to the data system.

8.3 LINEAR ION TRAP ANALYZER

The latest hot analyzer is the linear ion trap. It combines the separating capability of the quadrupole analyzer with the MS/MS capability of an ion trap. Trapping electrode rings are added to each end of the quadrupole rods to create the linear ion trap (Figure 8.9).

The analyzer can be run in a normal scanning quadrupole mode for separation and detection of mass ions, or the end electrodes can be turned on to retain a

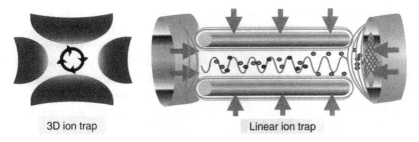

3D ion trap   Linear ion trap

FIGURE 8.9   Linear ion trap analyzer. (Courtesy of BioAnalytical Systems.)

specific ion in the trap for collision with a damping gas and fragmentation that may be aided with a supplemental resonance excitation voltage. The daughter ion fragments can then be released to the ion detector sequentially by scanning the dc/RF voltage on the quadrupole rods while utilizing a supplemental resonance ejection voltage on the trapping electrodes. The major advantage of a linear ion trap over a circular ion trap is the capacity of the linear ion trap. A normal ion trap is a point source trapping ions in a spherical segment between the ring electrodes. A linear trap spreads the sausage-shaped trapping volume down the center of the quadrupole pole rods, greatly increasing the trap's capacity. Reports in product brochures and the literature claim this increase to be 10- to 70-fold that of the circular ion trap. This translates to an increase in sensitivity for analyzing minor components of an HPLC effluent for trace analysis of metabolites or minor fragments from protein sequencing. Current linear ion traps are expensive, free-standing research instruments, but refinement and simplification of the technique seem to offer great potential for producing an inexpensive desktop LC/MS/MS.

8.4   TIME-OF-FLIGHT ANALYZER

The proteomics segment of the LC/MS system market (i.e., protein structure determination by LC/MS) are heavy users of time-of-flight mass analyzers because of their ability to analyze very large biological molecules. Time-of-flight LC/MS systems are commonly found in protein structure determination laboratories where LC/MALDI-TOF/MS systems are used for analysis of proteins, peptides, and polynucleotides. The liquid stream from the HPLC is collected on spotter plate wells, mixed with a chromaphore such as cyanocrotonic acid which will absorb light from the high-intensity laser burst in the mass spectrometer source. These target dye molecules explode, throwing the accompanying protein into the gaseous phase while ionizing it chemically. Since the free amino groups on side chains provide multiple ionization sites, multiply charged molecular ions are formed for the protein.

There are some questions raised as to whether this represents a true LC/MS system, since there is no direct injection from the HPLC system into the mass spectrometer. The interface consists of an automated spotter plate system that takes the HPLC effluent, mixes it with a sample matrix such as α-cyano-4-hydroxycinnamic acid, and air-dries the mixture. These spots are then blasted with a fast-response short-pulsed laser within the evacuated mass spectrometer injector for ionizing desorption into the flight tube. A time-of-flight mass spectrometer uses a UV laser to burst-ionize the sample from the GC in its source. The fragments are repelled down a flight tube through a focusing lens (Figure 8.10).

The flight time of each fragment is dependent on its m/z ratio, lighter fragments arriving at the detector first. To detect a given m/z fragment, each element of a diode-array detector is activated only for a given time-slice window, allowing selection of only a single mass per burst. Flight time is very rapid, on the order of 90 ns for a 2-m flight tube (Figure 8.11).

The timed array elements of the diode-array detector are set to sample the flight stream reaching the detector at different time windows. Using this technique, the entire burst fragment pattern can be analyzed for each event, greatly increasing the

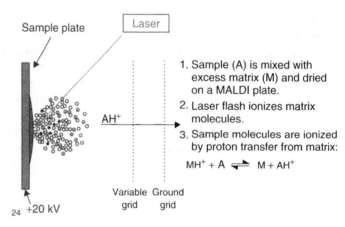

FIGURE 8.10   MALDI ionization source. (Courtesy of BioAnalytical Systems.)

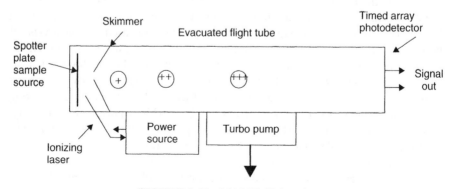

FIGURE 8.11   MALDI flight tube.

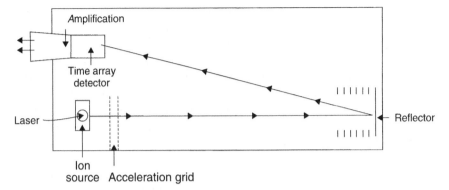

FIGURE 8.12 Reflectron TOF-MS.

sensitivity of the system. Summing the resulting time windows allows a 10,000-fold increase in sensitivity. Arrays are limited by the inherent noisiness of the array and the limited number of array elements available to do the sampling. A 50 × 50 array provides 2500 sample points. For the detection range 0 to 800 amu, this provides a resolution of 0.3 amu. Typical quadrupole resolution is 0.1 amu or better.

The length of the flight tube has historically produced very large, cumbersome TOF instruments. Folding the tube using electrical "mirrors" to reflect and accelerate the fragment flight stream back down the flight tube to strike the detector (Figure 8.12) has greatly reduced this problem. The fragments are repelled down the flight tube and separated at their m/z masses. Analysis of this family of molecular ions from a single protein that differ by the size of their charge, z, allows calculation of the molecular weight of the original protein. Charges as large as 20- to 50-fold on an ion radical allow enzymatic size proteins (MW = 25 to 60 kDa) to be separated on a time-of-flight mass spectrometer with a separating range of 0 to 2000 amu.

Although MALDI/TOF systems have dominated the time-of-flight market, newer systems using ES atmospheric interfaces and a holding cell before the time-of-flight analyzer are on the market as well as hybrid LC/ESI-quad/TOF/MS/MS systems in which the quadrupole analyzer provides specific ions for time-of-flight analysis.

8.5 FOURIER TRANSFORM ANALYZER DESIGN

Fourier transform mass spectrometry seems to offer interesting potential as a nondestructive LC/MS detector. It has lagged behind other LC/MS analyzers because it requires a fast, heavy-duty computer to make rapid on-line conversions of the large volumes of frequency to spectral data. LC/MS systems using a Fourier transform analyzer produce mass spectra using ion cyclotron resonance. The sample is ionized and solvent removed by an ion spray interface and passed into the analysis cell, where it is held in a constant magnetic field provided by a pair of trapping plates. Each fragment will follow a circular orbit with a cyclotronic frequency characteristic of its m/z value.

To detect the fragments present, a full-frequency "chirp" signal is applied from a pair of transmitter plates perpendicular to the trapping plates. Ions absorb energy from the chirp at their cyclotronic frequency and are promoted to a higher orbit. Detector plates perpendicular to the third plane of the cell measure a complex signal containing all the frequencies of the excited fragments (Figure 8.13).

Fourier transformation software converts this frequency snapshot to a spectrum of the m/z values present in the sample. Like spatial array detectors, every fragment is analyzed for every ionization burst event. Ion spray, electrospray, and laser-assisted ionization examples exist in the literature as input to Fourier

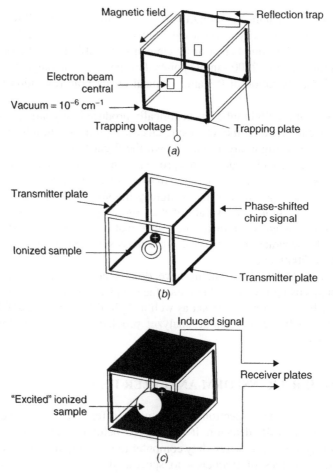

FIGURE 8.13   Fourier transform analyzer: (*a*) pulse; (*b*) chirp excitation; (*c*) resonate frequency signal.

transform analyzer LC/MS systems. The most common instrument using Fourier transform mass spectrometers are hybrid instruments using the Fourier transform analyzer to measure masses of daughter ions from a collision cell.

These analyzers offer a very desirable combination of measurement speed and sensitivity. Measurements can be made in milliseconds and have been used to monitor very short-gas-phase reactions. Since the ion fragments are not destroyed in the cell, multiple measurements over time can be averaged to produce a very accurate, high-resolution measurement yielding excellent sensitivity. The signal tends to be very stable and is not dependent on ion optics or variation in detector electronics. Modern computers can easily provide transformation calculation fast enough to provide real-time data.

8.6   MAGNETIC SECTOR ANALYZERS

The first research and commercially available mass spectrometry systems were magnetic sector instruments. They are not normally employed in LC/MS systems because of the slow response time of the sector instruments. But individual labs have connected magnetic sector analyzers designed for accurate mass measurements to HPLC with ion spray interfaces. The majority of the effluent is diverted to a secondary detector with only the sample of interest introduced into the sector source for accurate mass calculations to be used in determining C, H, N, O, and X ratios for technical structure determination. Magnetic sector analyzers use an electromagnet based on a large, massive permanent magnet to force ion fragments into circular sector flight patterns whose curvature is dependent on the fragments' mass/charge ratios (Figure 8.14).

The lighter the m/z mass, the more deflection it exhibits in the magnetic sector. Scanning of the mass range of ion fragments can be achieved in one of two ways, either by varying the accelerating voltage in the source or by scanning the electromagnetic field strength of the magnetic sector. Detection is done by using a moving slit and a photomultiplier tube or an electrooptical linear array detector.

The limitations of the magnetic sector systems are cost, the size and weight of the permanent magnet, response time, sensitivity, and linearity, especially at the high-mass side. The spectra obtained are generally not directly comparable to results from quadrupole or ion trap instruments for spectral extraction and library searching. Variation in the accelerating potential is limited in mass range, and sensitivity drops off at the high-mass end. Scanning the magnetic field is the more commonly used technique, but it suffers from reluctance, an inertia resistance to magnetic field change. This leads to slower scan rates, which translates to poorer sensitivity. Much of this sensitivity problem is overcome in modern instruments by employing spatial array detectors so that the entire mass range can be measured at all times, which increases the sampling rate and efficiency.

Magnetic sector instruments have made a comeback in the last few years because of their importance in accurate mass measurements for precise molecular-weight determination. Usually, injection is made from a probe rather than a

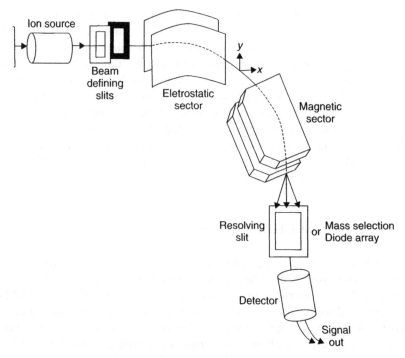

FIGURE 8.14   Magnet sector analyzer.

chromatograph. A technique called *peak matching* is used to compare the difference in accelerating voltage needed to make an unknown and a reference ion reach the detector in coincidence. The reference ion must be within 10% of the unknown compound mass, and masses can be measured to six-decimal-place accuracy with this technique.

Double-sector instruments are used to increase the precision with an electric sector in series to select ions of only one specific kinetic energy, and a magnetic sector to peak-match the reference and unknown compounds. Double sectors can be used for isotopic mass determination; they separate all ionic species into separate peaks that can be peak-matched against a reference compound.

9

MASS SPECTROMETER MAINTENANCE

The mass spectrometer's analyzer must operate in a very high-vacuum environment. Random collision with air molecules can disrupt the flow of ions from the source to the detector, causing signal loss. Ions that are introduced from an atmospheric interface must pass successfully through a pinhole orifice, possibly through a curtain gas into the source, where they are propelled into the analyzer by the repeller and focusing lens. The scanning dc/RF field of the analyzer sweeps the selected mass onto the detector surface. As the RF is swept higher, larger and larger masses impinge on the detector for measurement. In the end, these measurements are averaged at any point in time as a spectrum of m/z values versus signal strength. The total value of these signals are summed and plotted against time as a total ion chromatogram (TIC). The RF signal is then step down to begin the next SCAN.

9.1 HIGH-VACUUM OPERATION

Vacuums for mass spectrometry are established in two stages. A fore pump, usually an oil-sealed rotary laboratory vacuum pump, takes the vacuum down to 10^{-3} torr; then either an in-series oil diffusion or turbo molecular pump drops the analyzer pressure to 10^{-5} to 10^{-7} torr (Figure 9.1).

9.1.1 Vacuum Pump Types

The mechanical roughing or fore pump (Figure 9.1a) is an oil-sealed, rotary-vane vacuum pump commonly used as the laboratory workhorse vacuum pump.

LC/MS: A Practical User's Guide, by Marvin C. McMaster
Copyright © 2005 John Wiley & Sons, Inc.

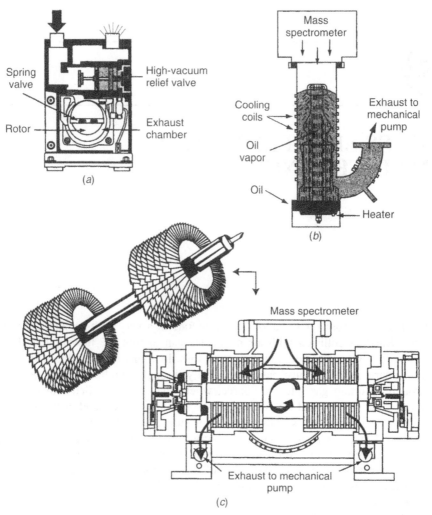

FIGURE 9.1 Mass spectrometer vacuum pumps: (a) rotary-vane vacuum pump; (b) oil diffusion pump; (c) turbo pump.

A piston riding on an eccentric cam attached to a drive shaft rotating in a compression chamber sealed by spring-loaded vanes moves gas from the inlet side to the exhaust port. It can reach 10^{-3}-torr vacuums only because of the vapor pressure of the sealing oil. Mechanical pumps typically exhibit pumping capacities of 50 to 150 L/min.

The oil diffusion pump (Figure 9.1b) sits between the inlet port of the roughing pump and the vacuum outlet of the mass spectrometer. Vacuums should be below 10^{-2} torr before the diffusion pump heater is turned on. Heated oil rises up the pump chimney, jets out through circular openings at various levels, condenses on contact with the cooled walls, trapping gases from the mass spectrometer, and runs down the sides, exhausting entrained gases into the roughing pump inlet. Diffusion pumps reach vacuums of 10^{-9} torr when chilled with liquid nitrogen. They can have capacities as high as 200 to 500 L/s, which can be important when pumping sources using high levels of gases for chemical ionization or when running ion-spray HPLC interfaces. Many systems with oil diffusion pumps have butterfly valves that snap shut in case of power loss to prevent contamination of the analyzer. This is an excellent feature if you are responsible for cleaning the analyzer. Although oil diffusion pumps provide excellent vacuums, they require maintenance on the pumping oil and bulky cooling systems. They are being largely replaced in modern laboratory mass spectrometer modules by the more compact turbo pump.

The turbo-molecular pump, commonly referred to as a turbo pump (Figure 9.1c), is like having a jet engine for your mass spectrometer. It has a series of vanes on a blade shaft rotating at speeds up to 60,000 rpm between an alternating series of slotted stator places. Air is grabbed by the blades, whipped through the stator slots, and then gabbed by the next blade. Only a small amount of air is moved by each turn of the shaft, but the number of blades and the high rotary speed move air rapidly from the analyzer chamber to the exhaust into the rough pump. Most turbo pumps have a dual set of vanes and stators on a single shaft feeding a dual exhaust. Turbo pumps come in a variety of sizes and are surprisingly efficient. The turbo pumps used to evacuate the ring electrode analyzer section of a desktop ion trap detector connect directly to the base of the analyzer section and are about the size of a double fist. Linear ion traps with a larger containment volume require a larger displacement system but still use a turbo pump to achieve their best vacuums.

It is important to vent a turbo pump to atmosphere before turning it off. Oil vapors can be sucked into a turbo pump from the roughing pump if it is left under vacuum. Some systems use all three pumps: roughing pump, connected to an oil diffusion pump on the source, connected to a turbo pump on the analyzer. In shutting these systems down, turn off the diffusion pump heater, allow it to cool below 100°C, vent the system, and then switch off the turbo pump. These differentially pumped systems are important if you are running an ion spray ionization source, where you will have very high source pressure.

9.1.2 Vacuum Measurement

Vacuum is measured in units of either torr or pascal. Torr, equal to the pressure of 1 mmHg or 133.32 pascal, is a commonly accepted vacuum measure in the United States. The pascal, equal to 7.5×10^{-3} torr (mmHg), is used more commonly in Europe.

Vacuum pressures are measured by two type of gauges. The medium-level vacuum of the roughing or fore pump can be measured by a thermoconductivity gauge such as a Pirani gauge. A heated wire is exposed in the vacuum line and is cooled by contact with molecules. The lower the contact rate, the lower the current draw and the lower the vacuum. High vacuums produced by oil diffusion or turbo pumps require use of a hot cathode gauge. Electrons streaming from the cathode are lost through contact with air molecules. The current produced is proportional to the concentration of air molecules present.

9.1.3 Pump Maintenance and Oil Changes

Roughing pumps are really the only pumps you will be expected to service. These should have their oil changed every six months. You can tell when it is time to change the oil by observing its color in the viewing port on the side. When it becomes brown and cloudy, it should be replaced.

Hoses coming from the roughing pumps should be checked periodically for cracks. Hose thickness and diameter are critical for proper performance of a vacuum system. You should avoid the temptation to substitute tubing of another size when you need to do emergency replacements.

Trained technicians from the manufacturer should service diffusion pumps. The same can be said about turbomechanical pumps. Most manufacturers offer some type of trading program for turbo pumps, and this definitely should be part of the purchase agreement when buying a system. When these systems go down, rebuilding them is a major undertaking. They operate at very high speeds with little tolerance for error. When they are down, they are down. It is important to have a replacement program to make sure that your LC/MS system is not out of operation until the pump is rebuilt and returned.

9.2 MS HARDWARE MAINTENANCE

The two major problems in day-to-day mass spectrometer operation, beyond reaching and holding a tune, are air leaks and solvent contamination or interference with low mass reading. Most of your attention will be devoted to the instrument's interface and source. Instrument manufacturers have all taken their own approach to MS design, the objective basically being to create ions and to calibrate and tune the source lens so that ions can be moved through the analyzer and reach the detector at the right place in the dc/RF scan. But no matter what instrument you have, you will have to clean the interface, source, and analyzer from time to time. How often depends on the rate of use, the nature of the samples you are analyzing, and the frequency specified by your protocol.

9.2.1 Problem Diagnostics

If you are having trouble reaching full vacuum but the filament will ignite, check for air leaks by scanning from m/z from 0 to 50 amu. Look for the water (18), nitrogen (28), and oxygen (32) peaks. If they are present, you probably have a leak around the interface to the source seal. Shut down the injecton port on the interface and the connection from the interface to the mass spectrometer source. If the fitting is snug, odds are that the ferrule is scored and needs to be replaced. Also check the seal around your cal gas valve if you have adapted a GC/MS instrument; I have seen occasional air leaks there. Often, opening the valve briefly and then reseating it can eliminate this source.

An excellent method of determining when you need to do source maintenance is to monitor the smallest calibration compound peak height from autotuning. Measure the height after tuning a new or freshly cleaned instrument. Set an acceptable threshold: say, 10% of the clean level. Once the target value drops below this minimum standard, it is time to clean the system. The calibration peak is chosen because larger peaks are much less easily affected by dirty analyzer surfaces or corroded source surfaces.

9.2.2 System Shutdown

Once your have decided to clean the source, you must vent the analyzer, power down, and cool the vacuum pump heater, if there is one. Next, remove the column and the interface from the mass spectrometer. The analyzer source assembly is removed from the analyzer and disassembled. The ion and filament contract surfaces must be cleaned and dried. The ion source assembly is reassembled and reinserted in the analyzer. The interface and column are reconnected, the analyzer reevacuated, and all temperature zones reheated. Finally, an autotuning is run to establish that the target peak height is back over the performance threshold.

The venting and power-down sequence will vary from instrument to instrument. It is important to follow exactly the procedure indicated in the instrument manual. Turbomolecular pumps need to be turned off at speed and evacuated through the roughing pump. Oil diffusion pumps must be cooled to less than 100°C before venting or they will stream oil back into the analyzer, contaminating the quadrupole surfaces. Once cooled, feel the temperature of the pump exhaust carefully; it is a pretty reliable guide that the diffusion pump is cool enough to turn off. Once the interface is removed, it is important to protect the interface surface until it is reunited with the analyzer section.

9.2.3 Source and Analyzer Cleaning

The mass spectrometer source on a LC/MS has fewer problems with source contamination than the GC/MS, where source contamination is caused by the filament cooking and charring hot vapors from the GC oven. But the LC/MS source must still be cleaned periodically, especially if the repeller lens is used to push the ionized sample from the interface or if the filament is used to initiate fragmentation, and the procedure is much the same. The LC/MS quadrupole analyzer is more susceptible to contamination by condensed sample deposits, especially if the ion spray interface is operated at a high HPLC flow rate. This will increase the need for analyzer washing.

To clean the source, you should first disassemble it in a clean area where you have plenty of room to work. Take special care not to lose small parts. Remove the control interface cables and the electrical connections to the filaments, the repeller, and the various focus lenses. Unfasten and remove retaining screws that hold the filaments and the repeller to the source body. On a Hewlett-Packard 5972 mass spectrometer, the entire source assembly up to the entrance lens can be removed as a single piece for disassembly and cleaning. The pieces that need cleaning are those in contact with the ion stream: the repeller face, the ion source inner body, both sides of the draw-out plate and its pinhole entrance, the focus lens, and the entrance lens contact surfaces. The ion source body shows burning next to where they contact the filaments. This needs to be removed and the holes leading into the source body need to be cleaned by reaming with a fine drill.

Cleaning source surfaces is an art and a source of controversy. A variety of abrasive, chemical, sonic, and electroplating techniques have been described in the literature. Abrasive techniques using fine powders are effective, require minimum equipment, and are reasonably safe for the source surfaces. Hewlett-Packard recommends cleaning surfaces with aluminum oxide powder and methanol to make a cleaning paste for use with Q-Tips in scrubbing these surfaces. They also supply aluminum oxide paper that can be used in cleaning inner surfaces, and drill bits and a holder for reaming out pinhole entrances. Some labs consider aluminum oxide too harsh and use less abrasive jeweler's rouge for cleaning. I am aware of one lab that uses nothing but a rouge paste that is intended for motorcycle detailing. Instrument service and repair facilities use high-pressure sand and water-blasting techniques to clean the source sent to them for cleaning and repair. Once the source elements are disassembled, you can also clean the flat lenses with jeweler's paste and a Dremel tool. Under no circumstances should you use abrasive stones or rubber wheels. These have the potential of scarring the surface, which can alter the electrostatic field of the lens. Once this occurs, that element may not perform as well as intended originally. Parts that are too small for the Dremel tool should be placed in a small beaker, and the larger flat pieces should be placed in a larger beaker. Fill the beakers with water and add a few drops of Alquinox (an industrial soap). Place the beakers in a sonicator, sonicate for approximately 1 hour, then remove the beakers and pour the water out very carefully (being especially careful not to lose any parts). Refill the beakers with methanol to remove the residual water. Sonicate again, for about 5 minutes. Remove the parts from the beakers and dispose of the methanol in an appropriate waste solvent container.

Other rinsing procedures call for sonication in a series of solvents. After wiping all surfaces with Q-Tips to remove as much abrasive as possible, they recommend sonicating the parts in a beaker for 5 minutes twice each in a chlorinated solvent such as methylene chloride, then acetone, and finally, methanol. They then air-dry, place all parts in a beaker, and dry in an oven at 100°C for 15 minutes. Be very careful not to clean Vespal surfaces abrasively or allow solvent to get under these surfaces. They tear up easily and solvent will cause them to swell, making reassembly difficult. It is usually easy to recognize these colored, plastic-looking surfaces.

If you are in doubt, don't clean the surfaces. Burn and char are usually pretty obvious. If you own a skillet, you are already familiar with them. Both types of char come from the same place, high-temperature oxidation of nitrogen-containing organic compounds. Atmospheric-pressure ionization produces relatively cool ions unless you are using a heated tube to finish removing solvents and volatiles. Generally, the filament is not turned on and does not add to the char problem.

Next, spread the parts out on lintless paper and inspect their cleanliness. At this point you should have on cotton gloves to prevent oils from your fingers from getting on the parts. If your source has small ceramic collars or spacers, inspect them for cracks or chips. It would be a good idea to replace these as necessary. You should maintain a good spare-parts inventory to cover this.

Another potential problem in a quadrupole that can affect its operation is accumulation of organics on the quadrupole rods. Ions that do not survive their travel through the quadrupole collide with the rods themselves, pick up an electron, and become electrically neutral molecules. If they are small volatile compounds, they are swept off the rod by the vacuum system and end as oil contamination in the vacuum pumps. However, there is a slow accumulation of larger, nonvolatile organics on the quadrupole rods. These must be washed off periodically since they will distort the electromagnetic field and eventually, shut down the analyzer. To remove the rods for cleaning, the system must be vented and the source removed as before. The ceramic collar between the source and the quadrupole is removed, electrical connections to the rods are removed, and the rod package is taken out of the source assembly. **The rods on most systems are held in exact hyperbolic alignment by two ceramic collars that must not be removed**. Nothing will shut a mass spectrometer down faster than messing up rod alignments. The minimum that must be done is to ship them in for repair and realignment. This is time consuming and expensive and not always successful.

The analyzer unit of four rods clamped in their ceramic collars is cleaned by sliding it carefully into a graduated cylinder and flooding with solvent. You must be very careful not to chip rods while placing them in the cylinder. Older quadrupoles have very large rod packages and are cleaned by wiping with large lintless paper towels. Usually, they are washed first with a nonpolar solvent such as hexane, then with methylene chloride, and finally, with dry acetone. Modern quadrupoles can be air-dried, then evacuated in a desiccator if the rod package is small enough to fit. Oven drying at 100°C has been rumored to cause rod distortion because of differential expansion of the collars and rods, and probably should be avoided. Wipe large rod packages and air-dry as well as possible, then finish the drying process as they are evacuated in the mass spectrometer. It won't help your roughing pump at all, but it will get the rods dry.

9.2.4 Detector Replacement

Electron multiplier detector horns have a finite lifespan and should be replaced when noise begins to increase. Run the repeller to its maximum value, then look at the electron multiplier (EM) voltage necessary to get a selected small calibration peak above the benchmark value. When the EM voltage exceeds 3500 V, it is time to consider replacing the detector.

9.2.5   System Reassembly and Startup

Reassembly is the reverse of the process of disassembly. Reassemble the source body, including the insulators, around the ion focus and entrance lenses. Insert the source body back into the analyzer body after reattaching the repeller and the filaments. Make sure that the ceramic collar between the source and the quadrupole does not bind—it must turn freely. Connect the repeller and filament leads as well as the leads to the focus lenses and the analyzer. When your source is assembled, you should conduct an electrical continuity check to ensure that, first, you have continuity where you are supposed to, and second, that no shorts exist. From here you are ready to reinsert your analyzer into the mass spectrometer body.

Once the interface is reinserted, connect it electrically to the LC/MS system controller. How you do this depends on the LC/MS design. Generally, you need to slide the interface fitting onto the mass spectrometer inlet. Tighten the fitting, turn on the vacuum pumps, establish the operating vacuum, and run a check for air leaks by looking for air and water peaks by scanning from 0 to 59 amu.

Replace the housing on the vacuum containment vessel. Turn on the roughing pump and begin evacuation. Set your interface heater to its operating temperature. When the gauge reads 10^{-3} torr, turn on the turbo or oil diffusion pump. For a diffusion pump, turn on the pump heater and bring it to temperature. Pump until you have reached your normal high vacuum, which will take about 2 to 4 hours on a Hewlett-Packard 5972 with a turbo pump, but will vary for other systems.

If there is a problem establishing the rough vacuum, push down the containment vessel lid. This is all that is necessary in most cases. If you still have a problem, stop the vacuum, inspect, and clean the gasket around the inside of the lid. If necessary, replace the gasket.

Once full vacuum is reached and you have checked for air leaks by scanning below 50 amu, you need to check the effectiveness of your cleaning procedure. Rerun you autotuning procedure and check the height of the target calibration peak. It should now be somewhere between your instrument's best measurement and your minimum system performance level.

9.3   SYSTEM ELECTRICAL GROUNDING

System electrical grounding is a problem that laboratories should never have. Grounding problems should have been worked out by the manufacturer before and during system installation and should not change until the system is altered. However, systems do wear, they are moved, and changes are made. When replacing controlling software and interfaces and moving to modern computers, grounding problems can occur.

For awhile I demonstrated replacement systems for a number of types of mass spectrometers. In a few cases, I saw problems with calibration peaks failing to stabilize. They jumped from side to side of the expected position. The problem disappeared only when the mass spectrometer chassis and the controlling interface were connected with a grounding strap and all electrical and computer systems power cord plugs were joined through a common surge protector. The problem seemed more common during the winter months, when laboratories are particularly dry. I suspect that static electricity discharges may be involved, since I have seen similar problem with other types of analytical systems during dry winter months.

A similar problem was seen and cured on a system that was shocked by a nearby lightning strike. This system had its original interface and computer system. Its calibration problem was very similar to that seen on demonstration systems and was cured by grounding the interface to the mass spectrometer. I did not encounter this problem often, but when it did occur, it was very frustrating. You could almost calibrate the system manually, but it would never complete an autotuning procedure sufficient for analytical studies.

10

APPLICATION AREAS FOR LC/MS

There are currently three principal application areas in LC/MS, but the technique has much wider potential application and is in fact already being applied to a variety of fields. The first main area is compound discovery and identity confirmation in pharmaceutical manufacturing or drug discovery. The second application area, called *proteomics*, is protein structure determination by LC/MS. A growing subset of these studies is in the field of DNA/RNA structure studies; although the name is not in common use currently, an appropriate term would be *nucleomics*. The third area of application is in metabolite and trace contaminates studies. Each of these application areas is covered in its own chapter because current interest in these areas is driving the growth of interest and sales of LC/MS systems.

10.1 COMPOUND DISCOVERY

Anyone doing organic synthesis or isolating compounds from biological sources has an interest, first, in establishing the activity of the compound, then its molecular weight, and finally, the compound's definitive structure. This is not simply to satisfy scientific curiosity. It is necessary before the synthesis or isolation of the compound can be published. It is an absolute requirement before the compound can be manufactured and sold to the public for treatment of a physical or mental problem: not only information on the compound of interest, but similar information on any compounds that may have been produced using the compound and might still be present when the compound was sold. It is critically important

LC/MS: A Practical User's Guide, by Marvin C. McMaster
Copyright © 2005 John Wiley & Sons, Inc.

to know the biological fate of the compound(s) when they are introduced into the environment and have had an opportunity to interact with other organic and biological entities.

There are three principal approaches to compound discovery: rationally guided targeted synthesis, activity assay-driven isolation from serendipitous synthesis, and activity assay-driven isolation of biologically active compounds. In the first of these, a compound is designed based on the structure of other known active compounds or on simple models of active biochemicals such as hormones or neural transmitters. A rational synthetic pathway is designed to produce the target compound from readily available precursor compounds. LC/MS can be used not only to confirm the structure of the final product and its impurities, but also to study precursor purity, intermediate compounds in the synthesis pathway, and the completeness of the conversion. HPLC has been used for years to monitor the steps in a synthesis. One of the first applications of HPLC was in aiding in the multistep total synthesis of vitamin B_{12} in Robert Woodward's laboratory at Harvard. LC/MS goes further in providing molecular-weight data for the compounds produced, and structure confirmation if needed.

Not all synthesis is this rational. Sometimes things are thrown into a pot, cooked up, and products are fished out of it to see if they are active. This is sort of the chemical equivalent of making chicken soup: not elegant, but something useful sometimes emerges. Obviously, you must have some method for rapid testing for some type of desired activity, whether that be as a headache remedy, a pesticide, or a treatment for AIDS. A positive result is then labeled as luck, serendipity, or good fortune; they all sell well once the LC/MS has provided the identity of the causative agent. Increasing activity against the screens drives the isolation of active compounds. LC/MS aids in the purification process and provides the identity of the most active compound or compounds. This is not an elegant process, but entire industries have been created out of it. Saccharine was recognized as a sweetener because a Swiss chemist took a break for a cigarette with laboratory chemicals on his hands and then went looking for the source of the intense sweet taste on his cigarette's filter. According to a seminar presentation by Karl Folkers, a former president of Syntex, the company grew out of studies done on a large block of cactus residue from tequilla production. Plant steroids in the mass proved to be starting materials for making the first steroid contraceptive drugs, and led eventually to the birth control pills used by many women today.

10.2 IDENTIFICATION OF COMPLEX BIOLOGICAL COMPOUNDS

The structures of three types of large biochemical molecules are of interest to biochemists and molecular biologists: structural proteins, antibodies, and enzymes; polynucleic acids, such as DNA and RNA; and complex polysaccharides that are structural and active components of cell membranes. These molecules are so large that while they can be separated and purified by HPLC or electrophoresis, it is very difficult to study their primary structure without first breaking them into smaller pieces by enzymatic cleavage. First the large secondary peptides, polysaccharides, or restriction fragments have to be separated, characterized, and identified. Common overlap areas are identified and a map of the original structure is reassembled from the structures of the pieces and knowledge of the specific cleavage points of the enzymes used. LC/TOF-MS has become the technique of choice, combined with two-dimensional gel electrophoresis for the separation and identification of the polypeptides, polysaccharides, and polynucleotide restriction fragments that are formed. On-line libraries of information and sequence structures of polypeptides are available to aid in quickly identifying the cleavage fragments. Research on polynucleotide restriction fragments and polysaccharides has lagged behind protein work. The very large sizes and rigidity of DNA and RNA fragments have made them difficult to separate above about 150-mer size. Polysaccharides have always been the stepchild of research. They were difficult to detect with most HPLC detectors, they had complex mixtures that were difficult to separate with most high-resolution columns, and they were not thought to make a major contribution to the life of the cell. The importance of polysaccharides and research on their structures by LC/MS has grown as more information has emerged as to their roles in membrane-mediated active metabolite transfer and as cell surface recognition sites in normal and cancer cells.

10.3 ANALYSIS OF TRACE IMPURITIES AND METABOLITES

The third large application area driving LC/MS growth is analysis of trace impurities and metabolites. Any compound that is synthesized or purified from cultural media brings with it other compounds that need to be analyzed, and usually removed, before doing activity studies, creating manufacturing processes, and creating formulations for distribution and sale. Government and regulatory agencies require that you know everything that is in the pot when you make a compound; everything, including formulation components, when the compound goes out the door; and everything that happens to these materials when they go into the environment for disposal. Some of these materials are present in only very tiny amounts, and the LC/MS systems required to analyze them must have very high sensitivity in the presence of large amounts of the target compound. Once the target compound is introduced into the environment, it is necessary to follow its breakdown products to determine its persistence and ultimate fate. I worked with a typical metabolite laboratory that had to analyze 3000 samples from soil, sludge, and river water from many different sites for a single product and do so in a three-month period. The metabolites were present in trace amounts and had to be either extracted or concentrated from a wide variety of matrixes and analyzed with very sensitive detectors. The LC/MS system combines high-resolution separating columns with a universal, high-sensitivity detector providing molecular-weight determination to aid in identifying traces of compounds. Trace metabolite analysis is an ideal area for application of hybrid LC/MS/MS systems to study fragmentation to help identify unknown components and to tie their structures to the original compound. LC/MS/MS allows us to ignore the main component while focusing on minor components. We can literally ignore the forest to see the trees—and not only see the trees but prove exactly what kinds of trees they are. Precolumn cleanup techniques applied early in the sample preparation can greatly aid sample simplification and concentration and extend our detection capability and simplify separation.

10.4   ARSON RESIDUE INVESTIGATION

Arson investigators look for signs of volatile accelerants such as gasoline and fuel oil in a fire site as proof of arson. In the past they have used GC/MS to identify the materials used to ignite a fire. Explosives can also leave residues since they are never completely consumed in the initial explosion. LC/MS uses known HPLC column techniques to separate explosives and mass spectrometry to identify the active compound definitively. Investigators can also identify the manufacturer by the deluant and formulation materials used to stabilize an explosive against accidental shock detonation. Many manufacturers are beginning to incorporate batch identifiers in their products to speed identification.

10.5   INDUSTRIAL WATER AND PESTICIDE ANALYSIS

Priority pollutants' testing is required in most states, but the exact nature and number of compounds required varies from one jurisdiction to another. Often, the compounds selected for analysis are chosen as much for our capability to analyze the target material as for its hazardous nature. Most compounds required for identification are volatiles and extractables suitable for GC/MS analysis and have methods mandated by the U.S. Environmental Protection Agency. Many more compounds used as pesticides and fungicides are known to be problems in the environment and are analyzed by HPLC and other techniques that do not identify them definitely. LC/MS methods have not yet been fully validated and accepted for their analysis. Mandated testing always follows when a need and ability to analyze such compounds are shown to exist. There is always a lag time between the introduction of new methodology and the approval of accepted and required methods from regulatory agencies. The rate of introduction will be increased by the availability of accepted HPLC analysis methods for many of these compounds. Adding the mass spectrometer as a molecular-weight detector is an easy add-on to these methods. Fragmentation information for quantitization and searchable library databases for structural confirmation will become available in a second phase of methods development.

10.6   TOXICOLOGY AND DRUGS OF ABUSE

Those detecting and identifying drugs were among the early adopters of GC/MS because it could provide a definitive identification of a particular drug or drug metabolite in blood or urine. The problem with the earlier technique was that the drug in question had to be thermally stable and volatile, or a volatile derivative had to be prepared from an extraction. Most drugs are taken by ingestion, inhalation, or injection and are usually water soluble. LC/MS has the advantage of being able to directly separate and analyze almost anything that is soluble. Fragmentation libraries exist to help in the identification of existing drugs of abuse. LC/MS/MS structural studies on new designer drugs can quickly identify their chemical nature. Cutting agents added to diluted street drugs can be identified and fingerprinted with LC/MS since most are sold in diluted forms. This information can be useful in helping to trace a batch of drugs up its chain of distribution and back to its supplier.

10.7 CLINICAL THERAPEUTIC DRUG SCREENING

One of the first applications of HPLC in a clinical environment was to titrate blood levels of theophylline, theobromine, and caffeine. Theophylline is an asthma drug with a narrow span between the therapeutic and toxic levels, the LD_{50} value of the drug. LC/MS used in clinical drug screening tells us the molecular weight of the compound we are examining, checks for trace impurities, and eliminates the problem of interference of a coeluting chromatographic compound with the amount of drug analyzed in the bloodstream. LC/MS/MS is also useful for following drug metabolites as an indicator of drug clearance rates in blood or urine.

A case study published on the Agilent Technology Web site describes large-volume screening of drugs of abuse using LC/MS at St. Olav Hospital in Trondheim, Norway. The total number of analyses is approaching 1 million per year on 24 LC/MS systems (Figure 10.1).

Traditional screening is done by immunology, which commonly identifies groups of compounds, has a high number of false negatives, and is high in

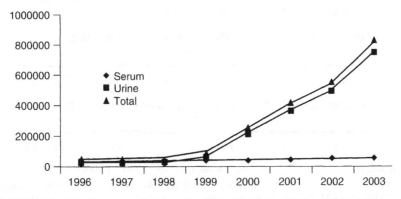

FIGURE 10.1 Number of analyses at St. Olav Hospital, 1996 to 2003. (Courtesy of Agilent.)

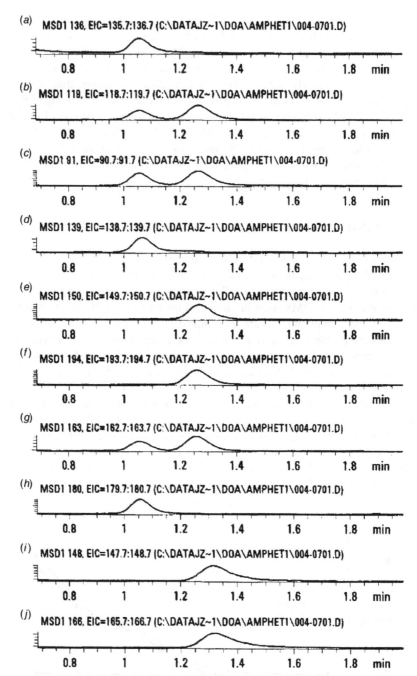

FIGURE 10.2 SIM amphetamine screening by ESI-LC/MS: (*a*) amphetamine; (*b, c*) amphetamine and methamphetamine qualifier ion; (*d*) ISTD; (*e*) methamphetamine; (*f*) MDMA; (*g*) MDMA and MDA qualifier ion; (*h*) MDA; (*i*) ephedrine; (*j*) ephedrine qualifier ion. (Courtesy of Agilent.)

cost. St. Olav's found that LC/MS was fast, quantitative, compound specific, and cost-effective. Both urine and plasma samples are screened, and as many as eight analysis are required for each test. Selective ion monitoring (SIM) is used to increase analysis sensitivity, and both ES and APC interfaces are needed to cover the broad range of drugs assayed. Each LC/MS unit is configured in the same way, with a quaternary mobile-phase system with methanol, acetonitrile, ammonium acetate, and formic acid, and is used for both drugs of abuse (DOA) and therapeutic drug monitoring (TDM). Two short columns are needed for DOA screening, a C18 and a CN. GC/MS is used for confirmation because it is still the "gold standard" for legal technique, but comparison of results, shows close to a 100% accordance with LC/MS results, with no false positives. An example is provided of a series of amphetamine panels used in these studies run on a CN column with an ES interface with d^3-amphetamine as an internal standard (Figure 10.2).

10.8 PESTICIDE MANUFACTURING

Large-volume manufacturing of pesticides can be looked at as a model for most industrial chemical manufacturing of compounds such as surfactants, optical brighteners, solvents, plasticizers, or dyes. Traditional analysis has been by HPLC and gas chromatography, with GC/MS playing a larger and larger role. Herbicides, insecticides, fungicides, and plant growth regulators have to be characterized before going into production and into formulations for distribution. LC/MS is starting to come into analytical laboratories for these molecules, which are not easily characterized or quantitated by gas chromatographic analysis. LC/MS is important in impurity studies, in monitoring process changes, and in diagnosing the nature of problems when things go wrong in the process and in recovering product from accidental loss.

I have been involved in monitoring a 6000-gallon herbicide production unit when an operator on the third shift messed up a product washing operation and product had to be recovered from the sump where he dumped it accidentally. I have assisted in analyzing recovery of product from off-spec production runs so that it could be formulated and sold. I have helped run a study trying to determine where 20% of 95,000 pound/year herbicide production was lost during processing and appeared later to poison the bacterial effluent treatment system. All of these involved unknown compounds that would have been analyzed easily and ideally using LC/MS and LC/MS/MS systems, had they been available.

Hazmat studies of industrial waste dumps and Superfund sites involve large nonvolatile compounds. It is important to characterize the organic compounds present, their persistence in the soil and environmental waters, and environmental metabolites to aid in their cleanup and to trace them back to their points of origin. LC/MS is the ideal equipment for providing molecular weights and definitive structure characteristics of these hazardous materials.

11

TRACE ANALYSIS AND LC/MS/MS

Often, retention times and molecular-weight information from a mass spectrometer's total ion chromatogram are insufficient to identify completely all the compounds present in the effluent from the liquid chromatograph. Trace impurities require greater sensitivity than can be provided by a mass range scan. At times, we need more information about the target molecule than can be provided by a simple molecular ion scan. At this point we need to turn to an LC/MS/MS system and collision-induced fragmentation of the molecular ion to produce daughter ions resulting from fragment-induced cleavage and rearrangements resulting in the loss of neutral molecules. We can estimate the structures of these daughter ions by examining the mass intervals and isotopic patterns between the product ions. But this requires use of more complex and more expensive LC/MS equipment.

11.1 LC/MS/MS TRIPLE-QUADRUPOLE SYSTEM

The electron impact (EI) mode of a GC/MS system run at 70 eV ionization potential provides enough energy to break the parent molecular ion into various fragments characteristic of the parent molecule. Fragment analysis can provide information on the structure of the original molecule that is missing in the normal TIC and spectrum resulting from the low-energy ionization techniques used in LC/MS. Examination of the mass spacing between fragment ions and peak combinations allowed us to determine missing masses, the presence of

LC/MS: A Practical User's Guide, by Marvin C. McMaster
Copyright © 2005 John Wiley & Sons, Inc.

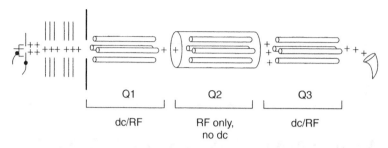

FIGURE 11.1   Triple-quadrupole LC/MSMS system.

halogen, nitrogen, and sulfur atoms, and to estimate the molecular weight if the molecular ion is missing. Searchable computer libraries of fragments and their corresponding structures have been built up to allow determination of a compound's structure. Obviously, similar types of structure information would be useful to an LC/MS investigator.

The triple-quad LC/MS/MS system was designed to cleave ions into their daughter ions. It consists of a scanning Q1 quadrupole analyzer for separating the original precursor ion(s), an unscanned Q2 quadrupole that serves as a collision cell to fragment the ions sent to it by collision with a heavy gas molecule, and a scanning Q3 quadrupole that can separate the fragments produced in the Q2 unit (Figure 11.1).

11.2   MS/MS OPERATING MODES

The first quadrupole, Q1, is operated in either a full-scan or SIM mode to select ions to pass on to the other analyzers in the system. The middle Q2 unit is flooded with a heavy inert gas, either krypton or xenon, and fragmentation is induced as the ions passed to Q2 from the first quadrupole undergo thermal collision with the higher concentration of large molecules. The final analyzer, Q3, can also be selected for either full-scan or SIM mode operation. We have two operational modes for both Q1 and Q3, providing four possible experiments that are run with a triple-quad mass spectrometer (Figure 11.2).

There are four possible modes of operation of the two analyzers: Q1 scan/Q3 SIM, called *daughter mode* or *precursor scanning*; Q1 SIM/Q3 scan, called *parent mode* or *product scan*; Q1 scan/Q3 scan, referred to as *neutral loss scanning mode*; and Q1 SIM/Q3 SIM, referred to as *multiple reaction monitoring* (MRM) *mode*.

SCAN/SIM mode operation lets us determine which primary fragments are related to each other. The first quadrupole is scanned over the mass range, and all fragments formed enter the collision cell and fragment to form secondary fragments. The third quadruple is parked at a specific mass/charge position and only primary fragments that break down to form this specific secondary *m/z* value will be detected. This common daughter ion points out interrelated primary fragments

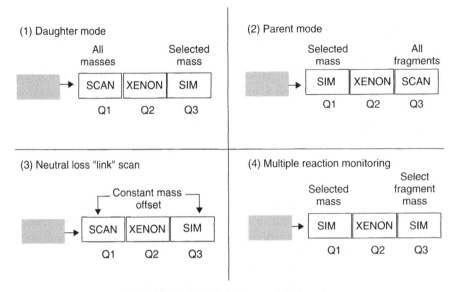

FIGURE 11.2   MS/MS operational modes.

and helps us to understand which fragments are formed when a large primary fragment breaks down. In LC/MS we could store a TIC run of a peptide mixture for retention times and molecular-weight determination and then run Q3 parked at mass 79 and lay this chromatogram over the TIC to detect phosphorylated peptides present in the mixture.

SIM/scan operations parks the first quadrupole analyzer at a specific mass, allowing only a single primary fragment to enter the collision cell where it fragments into secondary ions. The final quadrupole is run in full-scan mode, detecting all secondary product fragments formed from this single primary parent, again providing structural information by showing its breakdown products. This mode is commonly used in LC/MS to examine a molecular ion for its mass spectrometer fragments to provide structural identification.

Scan/scan operation, or neutral loss mode, is a little more complicated since both analyzer quadrupoles will be scanned at the same time but with a preset mass offset. When a primary fragment undergoes further fragmentation, it breaks into two pieces, a charged secondary fragment and a neutral molecule. What we are detecting in this mode are primaries that lose the same neutral molecule, such as carbon monoxide, water, or a vinyl compound, and therefore may be breaking down by the same fragmentation mode. The molecular mass of our suspected "neutral loss" is the value we assign to our scan offset between the two scanning quadrupole analyzers. All primary fragments separated in the first analyzer enter the collision cell and fragment. Only secondary fragments whose mass is exactly the neutral loss smaller than their primary fragment are detected after the final quadrupole and show up in the display of the scanning chromatogram. Any primary fragment that breaks down by forming a neutral molecule that has a mass different from the offset mass will not be selected by the second analyzer and enter the detector.

SIM/SIM operation is designed to analyze specific components of very impure mixtures definitively without having to purify them completely. They can be detected at a very high sensitivity since both analyzers are parked at different specific single m/z values, and a greater number of scans can be summed in determining their position. Nature makes very complex mixtures that cannot always be separated completely either though extractions or by chromatography. We examine a chromatographic peak in which we expect a specific compound to appear by using the first quadrupole to separate a primary fragment characteristic of the compound of interest, pass it into the collision cell, and use the final quadrupole to identify it by looking for only one of its specific daughter ions. We can identify and quantitate each targeted compound in a mixture, even if the chromatographic peaks that contain them are contaminated. For each compound to be analyzed, we select an individual primary and secondary fragment on a time basis, in step with their expected chromatographic retention time.

Applied BioSystems provides an illustration of a metabolite study of gly-buride in their *Drug Discovery and Development Bulletin*, Issue 1, Summer 2004: Metabolite ID Edition. Glyburide is an oral sulfonylurea drug used to treat type-2 diabetes. An LC/linear ion trap system proves the existence of glyburide plus five major and five minor metabolites in the total ion chromatogram (Figure 11.3).

Precursor neutral loss (125 offset) scanning of glyburide and the major metabo-lites provide information needed to identify each metabolite by using product ion isotope patterns to show chlorine loss and oxidation. The remaining trace metabo-lites are identified by the increased sensitivity gain using SIM/SIM-based MRM studies as shown in the table in Figure 11.3.

11.3   ION TRAP MS/MS OPERATION

As mentioned earlier, ion trap mass analyzers have an inherent MS/MS capability. The ring electrodes on the classical three-dimensional ion trap store all the ions that it receives, and then releases them as the dc/RF voltage is increased in either a scanning mode or in a stepwise SIM fashion. The ITMS ring electrode is equipped with a gas inlet port so that the trapping volume can be flooded with an inert gas such as carbon dioxide to stabilize the ion holding pattern. The destabilizing frequency can be raised and lowered to release all other ions to the detector while selectively retaining only a single-ion mass range. Replacing the carbon dioxide with a heavy inert gas such as xenon allows the trapped ions to undergo fragmentation collision to form daughter ions that can be held and then destabilized to release them to the detector by changing the trapping frequency. This can be done in a full-scan mode to produce a daughter mode scan or in a step fashion to product a single daughter ion.

A linear ion trap (LIT) can also be used to produce MS/MS scans. It has the advantage of being capable of storing many more trapped precursor ions. The

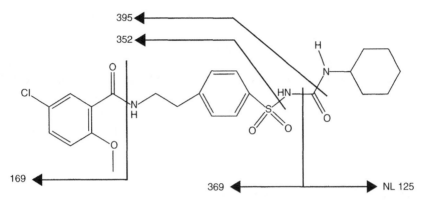

Glyburide is a sulfonylurea oral drug used to contol hyperglycemia in type 2 diabetes mellitus.

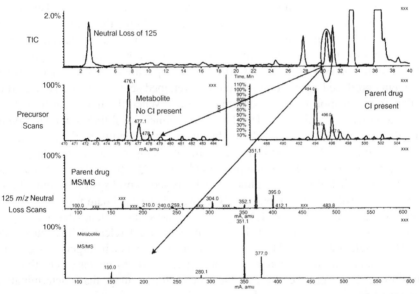

Cl loss plus oxidation, isotope pattern, and MS/MS confirmation found by neutral loss of 125.

Metabolite—Parent Drug (494)	EMS	Precursor Ion	Neutral Loss	MRM
Oxidation (510)	5	5	1	5
Di-oxidation (526)	0	3	0	4
Dehydrogenation (492)	1	0	0	▲1
Loss Cl + Oxidation (476)	0	0	1	NA
Ring Loss + Hydrogenation (414)	0	0	1	NA
Ring Loss + Oxidation (427)	1	1	1	NA
Ring Loss (412)	1	1	1	NA
Total Metabolites Found	8	10	5	10

Summary of metabolites found with the different IDA survey modes available on the 4000 Q TRAP system.

FIGURE 11.3   A/B glyburide metabolite study. (Courtesy of BioAnalytical Systems.)

three-dimensional trap is limited to its spherical-segment volume, but the linear ion trap can spread the trapped ions in the full volume trapped by its quadruple rods. Increasing the number of trapped ions before subjecting them to heavy ion fragmentation collision increases the signal strength when the fragment ions are released to the detector either by a scanning or a SIM destabilization of the dc/RF signal. This signal allows the LIT to be used to increase the detectability of trace contaminates, or it can be used to increase the precision of measurement of the m/z signal, allowing accurate isotopic masses to aid in precursor accurate mass structure determination.

11.4 HYBRID LC/MS/MS SYSTEMS

The success of the triple-quadrupole MS/MS system suggested immediately that other combinations of two mass analyzers connected on either side of a collision cell would be advantageous. The first attempt was to build a quadrupole/three-dimensional ion trap MS/MS combination, which had the advantage of providing its own collision cell. It also allowed the first application of MS^n, in which ion fragments produced in the ion trap could be retained selectively and subjected to further fragmentation to aid in structure studies. When the linear ion trap (LIT) began to emerge, it quickly replaced the three-dimensional trap in preparation of a LC/Qtrap system because of the higher trapping capability of the quadrupole-based ion trap. It also offered the capability of being run as a true QqQ triple quadrupole as well as in its trapping mode as a Qtrap. It also allowed MS^n operation like that of the three-dimensionally based hybrid system (Figure 11.4).

The time-of-flight (TOF) mass analyzer was also a candidate for hybridization with a quadrupole. In the Q/TOF type of LC/MS/MS system, the quadrupole selects ions that are held in an ion accumulator until released in a burst into the TOF flight tube. The TOF analyzer can analyze much larger masses, and the accumulator allows accumulation of a specific ion to increase sensitivity or allow much more precise mass measurements for accurate mass determination (Figure 11.5).

Hybrids of the quadrupole/Fourier transform mass spectrometer (LC/QFTMS) are beginning to appear on the commercial market. These are designed to take advantage of the extremely high sensitivity inherent in the nondestructive FTMS

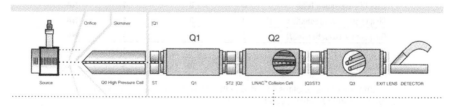

FIGURE 11.4 Qtrap hybrid LC/MS/MS system. (Courtesy of BioAnalytical Systems.)

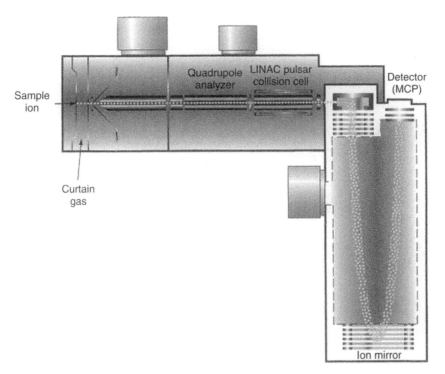

FIGURE 11.5   QTOF hybrid LC/MS/MS system. (Courtesy of BioAnalytical Systems.)

analyzer, which can hold ions in its cell and excite them repeatedly with its chirping frequency to increase signal accuracy for metabolite and accurate mass studies. It also is a very fast analyzer when equipped with modern computer-transforming software and is being used to analyze and follow reaction changes when used with ultrafast HPLC systems and columns.

Obviously, other hybrid analyzer systems are possible, and someone will probably make one. MALDI-TOF/TOF systems have been created to aid in protein and peptide structural analysis and probably will have some application in a combination LC/ESI-QTOF/TOF system for protein structure determination.

12

DRUG DISCOVERY AND BENCHTOP LC/MS

New drug discovery has been one of the two major applications that drive the growing market for new LC/MS systems. The estimates are that 100,000 candidate compounds must be screened for every new drug that reaches consumers. Each of these compounds must be synthesized, at least partially purified, and when it shows activity it must be identified, purified, and characterized before it can go into preclinical and clinical testing. Combinational chemistry and rapid, standardized LC/MS chromatography and identification have been the workhorse tools used to provide the drug candidates to fill this massive pharmaceutical pipeline. A candidate drug then moves into animal testing, clinical testing on humans for FDA approval, and on into manufacturing and formulation. Throughout the process, LC/MS provides rapid, sensitive separation and determination of retention times, molecular weights, and structures of by-products, metabolites, degradation products, and artifacts from packaging and formulation (Table 12.1).

12.1 ACTIVITY SCREENING

The first requirement for rapid screening of a large group of compounds is some method of evaluating activity. You must have a therapeutic testing target or a problem to be treated. This can be anything from the enhancement of a metabolic function to a treatment to block a disease causative agent such as the HIV virus. Initial testing must be done in a rapid-response bioassay or in vitro matrix so that a large number of responses can be measured in a very short time.

Compound activity screening is not new, only the response times required. When I was in graduate school we were required to submit at least 3 g of any

LC/MS: A Practical User's Guide, by Marvin C. McMaster
Copyright © 2005 John Wiley & Sons, Inc.

TABLE 12.1 LC/MS in Drug Discovery

Stage	Candidates	LC/MS System	Application
Drug discovery	100,000	LC/MS	Screening
		LC/MS	Purification
		LC/MS/MS	Identification
Preclinical	100	LC/MS	Bioassay monitoring
Clinical	10	LC/MS	Blood/urine assay
		LC/MS/MS	Trace metabolites
FDA approval	1	LC/MS/MS	Biological fate, degradation products
Manufacturing	1	LC/MS	Process monitoring
		LC/MS/MS	Reaction by-products
Quality assurance/ quality control	1	LC/MS/MS	Formulation artifacts
		LC/MS	Process troubleshooting

compound we prepared. This was placed in a storage bank in glass vials, and samples were withdrawn for testing in a number of commercial activity screens by a variety of companies. Later, as a synthetic chemist with the Industrial and Biochemicals Department of E.I. DuPont de Nemours, I prepared a series of compounds that were submitted to a variety of screens, evaluating candidates for pharmaceuticals, insecticides, fungicides, herbicides, and plant growth stimulants. I was assigned to the insecticide and herbicide screening evaluation committees for awhile, and we saw compounds that were prepared by other departments, intended as polymer precursors, optical brighteners, and pharmaceutical intermediates as well as pesticides. Any compound prepared in the company and in many of the local universities were submitted to all the company evaluation screens. One of the most promising inhibitors for grass growth that we looked at was a precursor for vitamin C. You sprayed it on your grass once a month and the grass stayed green and exactly the same length the entire summer. Unfortunately, after one more oxidation step, the product became vitamin C and could be sold for 1000 times more than you could ever get for it as a grass growth inhibitor.

Most compounds submitted to screens are from one of two sources. *Rational synthesis* designs a specific compound synthesis to produce a single compound of known structure. Generally, this is not a single-step synthesis, so there are precursor compounds produced on the way to the final target compound. All of these compounds were designed rationally, but they probably were not designed for the activity they may show in an activity screen. All are submitted to activity screens for testing against a variety of target bioassays. Activity is the result of serendipity or blind luck. You try and build on that luck if you happen to get an active response. Usually, the first compound showing activity is not the most active form of that type of compound. Making analogs with trivial structure changes can sometimes produce a relative with greatly enhanced activity. So you try to design changes that can increase or decrease solubility, volatility, hydrogen bonding, and polar or nonpolar characteristics. You replace a methyl side chain with an isopropyl group, oxidize an aldehyde into an organic acid, and then convert it into a series of esters. Using this approach to synthesis, a chemist could generate perhaps 50 new compounds a quarter for the activity screens.

Combinatorial synthesis takes a different approach to generating the large volume of compounds needed to fill the pharmaceutical pipeline. Serendipity is driving the nature of the actual drug candidates and is simply dependent on the total number of compounds for the biological screen to select among. In rational design, synthesis increasing the number of compounds generated or cloning around an active compound is dependent on the number of chemists available and how hard they can be pushed to generate new compounds. A new approach was needed to move this volume to a new order of magnitude. The combinatorial approach employs parallel synthesis of multiple compounds in a single reactor rather than slower direct sequential synthesis of analogs. It relies on the resolving power of LC/MS to separate and identify the molecular weights of the components in the combinatorial library that is generated.

Combinatorial synthesis arose out of a modification of the Merrifield synthesis for preparing synthetic peptides and the Edman degradation procedure for sequencing peptides. In the Merrifield synthesis apparatus the first C-terminal amino acid of the peptide desired is attached to a polystyrene bead by way of a chemical linker molecule. Subsequent amino acids are added at the N-terminal end of the growing attached peptide. Cleaving the linker and filtering the beads releases the final target peptide or protein molecule. In the Merrifield synthesis, the reaction connecting a new reagent amino acid to the attached peptide must be driven to completion at each step to ensure formation of only a single product peptide.

A mixture of peptides is formed if only partial amide formation occurs at the N-terminal amino acid at any step in the synthesis. You are faced with a very difficult purification process when the final peptide mixture is removed from the bead's linker. If we were to add a mixture of 20 reagent amino acids in the first reaction step with the attached C-terminal amino acids, we would obviously form 20 bound dipeptides having the same C-terminal amino acids.

Combinatorial chemistry aims at building a library of compounds in a manner similar to the dipeptide mixture. The starting material for a target drug is attached to the bead's linker arm and a synthesis sequence is carried out to build the core molecule. In the last synthesis step, a reactive functional group on the bound-core drug candidate is treated with a mixture of reagents to build our combination of analogs of the core molecule. These can be released by cleaving the linker and filtering the beads to produce our combinatorial library in solution ready for LC/MS purification and analysis. The entire library or just compounds purified by LC/MS can then be taken to the bioassay for evaluation. The mixture can be separated and individual compounds assayed if any compound in the library responds to the bioassay.

12.2   STANDARDIZED LC/MS SCREENING

One of the key components for rapid screening was the creation of a standard or generic LC/MS methodology used throughout the company and for all stages of product evaluation and development. No attempt is made to optimize the chromatographic separation for individual compounds or classes of compounds. A standard 5- to 20-minute linear reverse-phase gradient from 5% organic in water to 95% organic at neutral pH is run on a specific C18 column type from a single manufacturer. The mass spectrometer is run in scan mode over a standard *m/z* range with a time delay to ignore the solvent spike and retention time, and molecular ion mass is determined for each compound.

This technique is similar to the HPLC scouting gradient I used in laboratory demonstrations to prove separation feasibility. I always ran the same first gradient regardless of the compound and was always able to resolve something on the first injection that would guide rapid development of an optimized analysis.

The standard LC/MS assay technique is used from drug development screening through metabolite and degradation studies on compounds from almost any media, such as urine, plasma, or reaction mixtures; only the sample preparation will vary. The advantage of generic run conditions is that it allows preparation of an LC/MS separation database that can be referenced for compound mixtures from anywhere in the development and manufacturing process cycle. It trades off resolution for consistency, speed, and a decrease in method development times. It permits creation of a computer-searchable database of information for all of the compounds being investigated in a company. The mass spectrometer provides sensitivity and resolution gain as well as information on retention times and molecular weights. Further structural and trace metabolite information can be obtained by running the same method on an MS/MS system and using target compound template information from the precursor scan to interpret product ion scans.

When I first saw the linking of combinatorial chemistry with generic LC/MS methods I was appalled to the depth of my traditional chemist's soul. It looked like they were using the power of the mass spectrometer resolution to try to fix bad chemistry and bad chromatography. But when I studied the technique, I realized that it used the same resolution, speed, and load triangle that we take advantage of in doing preparative chromatography. There we sacrifice resolution to gain load and speed. In generic chromatography they trade off resolution for speed and analytical compatibility across the drug discovery and development process.

There may be one exception to the use of generic chromatography. Many drugs are water-soluble basic compounds that are poorly retained on reverse-phase columns and may tail badly on poorly end-capped columns because of ion-exchange interaction with the silica support material. Much of this problem can be overcome by using the new generation of hybrid columns and by using volatile buffers to control pH at 10.0. The chromatographic separation is better, but additional problems are introduced. Data produced under nonstandard conditions may distort the information in the database. The volatile buffers may create ammonia or triethylamine adducts that will complicate or confuse molecular-weight determinations and may also interfere with bioassays. Generally, it is better to select standard reverse-phase columns that have been well end-capped with protected surfaces and use the mass spectrometer's sensitivity to overcome resolution problems. Starting gradients at very low organic levels should allow retention of almost any drug candidate.

12.3   MOLECULAR FRAGMENTATION FOR STRUCTURAL DETERMINATION

Retention times and molecular-weight determinations suffice for screening a drug candidate until it begins to show activity in bioassay screening. At that point it is necessary to purify the compound and characterize its structure before moving on into clinical screening and Food and Drug Administration registration. The standard LC/MS run used to examine the combinatorial library can be optimized for separating the active compound from other candidates and trace impurities in a preparative purification. Running these optimized conditions in a LC/MS/MS system lets us use additional mass spectral modes to gain further information on the drug candidate.

The first thing that would be done is to isolate the molecular ion of the peak of interest, carry out collision fragmentation in the present of heavy gas molecules, and then run a precursor scan (Figure 12.1). The fragmentation pattern intervals can be analyzed for structural determination, or the fragment scan can be submitted for a library search of known structures. The precursor fragment scan can also be used as a template for additional study of the structure. Selected fragment ions can be further fragmented and analyzed by product ion scanning. Referring these daughter ion scans back to the precursor template allows us to determine the steps involved in the original fragmentation and hopefully, leads us to the original structure. The MS/MS operation can also provide information on loss of neutral molecules during fragmentation and lead to a study of the formation of specific fragments from other fragments using MRM (SIM/SIM) studies.

The product scan template can also be used to study reaction impurities, metabolites, and degradation products of the target compound. Metabolites retain core structures present in the target molecule. Metabolite precursor and degradation product scans will produce some of the same fragments as those found in target molecule precursor scans and thus aid in tying the metabolite to the original compound. Impurities are often related to the core of the target drug, and comparison of precursor scans of the impurities to the target drug precursor

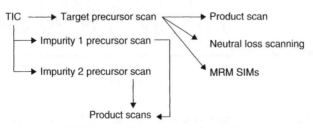

FIGURE 12.1   Fragmentation study.

template can help in identifying impurities' structures. Structures of impurities, metabolites, and degradation products will all be needed to control manufacturing and in submitting a target compound for drug registration approval.

12.4 PROCESS MONITORING

The drug manufacturing process is also monitored using the standard LC/MS chromatogram generated during drug discovery and preclinical development.

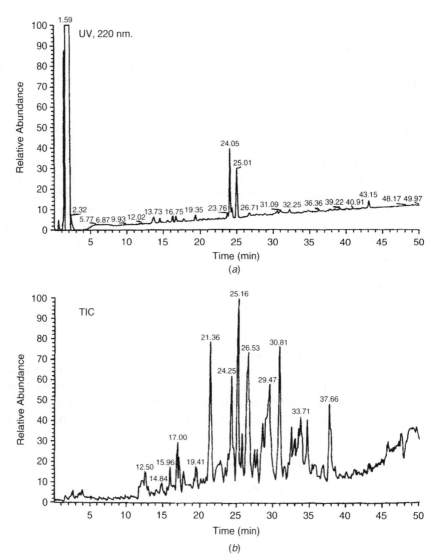

FIGURE 12.2 Chromatograms of adhesive extract: (*a*) UV chromatogram at 220 nm; (*b*) TIC full-scan MS chromatogram. (From Tiller et al., 1997.)

When abnormalities develop in the process reactor, quality control information on impurities, metabolites, and degradation products can be used to troubleshoot and help correct the problem. Formulation and packaging of the finished drug can introduce additional complications and compounds, and these also need to be monitored.

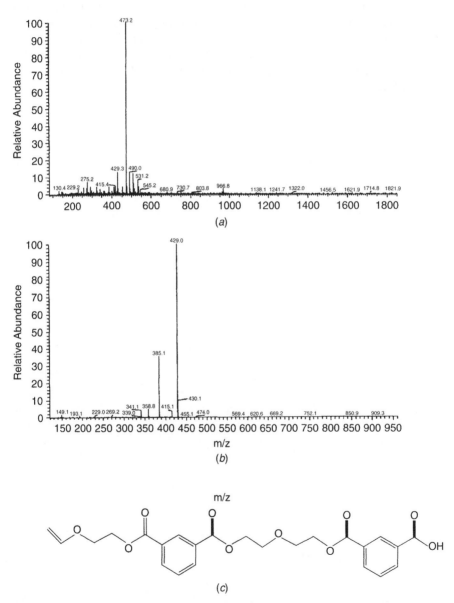

FIGURE 12.3 Mass spectra and structure of adhesive component: (*a*) full-scan mass spectrum of component; (*b*) product ion spectrum. (From Tiller et al., 1977.)

Tiller et al., have published a very instructive study of the extraction of an adhesive used in drug packing that is summarized by Mike Lee on pages 174–175 of his book, *LC/MS Applications in Drug Development*. The extracted mixture is injected and monitored on both an ultraviolet detector at 220 nm and a mass spectrometer as the total ion chromatogram shown in Figure 12.2. This comparison illustrates the difference in resolving power of the two types of detectors. The MS peak at 25.16 minutes is then subjected to fragmentation and a precursor scan (Figure 12.3a) in the ESI-LC/ITMS system to yield the [M + H] ion with a molecular weight of 473.2. A product ion scan (Figure 12.3b) of this 473 ion provides two major ions. The presence of the 429 and 385 ions in the product spectra correspond to two successive losses of either C_2H_4O or CO_2, suggesting the structure proposed in Figure 12.3c.

13

PROTEOMICS: LC/MALDI/TOF AND MS/MS LIBRARIES

Proteomics is the name applied to studies of proteins, peptides, and amino acids. Proteins are built up in the cell from 23 amino acids on a template provided by messenger RNA. The protein formed can then be changed by the addition of molecules such as lipids, phospholipids, and complex polysaccharide molecules to form posttranslationally modified (PTM) proteins.

Proteins are large, complex, three-dimensional molecules that serve as structural components of a cell combined with lipids, as supports for active molecules such as hemoglobin and antibodies, and serve as biological catalysts as enzymes and hormones. Protein molecular weights typically vary from 40,000 to 80,000 Da for enzymes to hundreds of thousands of daltons for antibodies and structural proteins. Peptides, many of which act as hormones, are made up of smaller sequences of amino acids and may have molecular weights from less than 1000 Da up to those of small proteins.

In understanding proteins it is necessary to understand three protein structure types. The primary structure is a protein's amino acid sequence. The secondary structure is the three-dimensional folding of a protein under the influence of hydrogen bonds and disulfide cross-linking and is often critical for enzymatic activity. Tertiary structures result from interaction of layers of multiple protein molecules sometimes mixed with other molecules, such as lipids.

Proteins are usually found in complex biological mixtures, and a wide variety of them exist. The majority of proteins are housekeeping proteins involved with maintaining cells and organs or transporting metabolites needed to feed the cells, such as albumin and heme proteins. The protein of interest is usually present in

LC/MS: A Practical User's Guide, by Marvin C. McMaster
Copyright © 2005 John Wiley & Sons, Inc.

only minute amounts and must be isolated laboriously before its structure can be studied. HPLC gel filtration was very instrumental in reducing the time needed to purify proteins while maintaining enzymatic activity.

The goal of much of the growth of LC/MS system purchases in universities is rapid isolation, purification, and identification of the structures of proteins and PTM proteins. In the past, once a protein was isolated, identifying the protein structure could be a long, involved process of enzymatic degradation of the protein followed by laborious sequencing of the product peptides and then slowly rebuilding of the original structure from knowledge of the parts. Initial sequences of simple proteins took years until automated sequencing cut the process to weeks. A combination of two-dimensional gel electrophoresis to separate proteins with LC/MS separation and identification of the enzymatically cleaved peptides in the scrapped and extracted gel spots has cut protein structural analysis to hours. This has been aided greatly by building computer-searchable database libraries of peptide mass spectrometer fragmentation patterns.

13.1 PROTEIN MOLECULAR-WEIGHT DETERMINATION BY LC/MS

The first step in determining protein structure is to determine the molecular weight. We can get a rough idea of the molecular weight during HPLC separation using a gel permeation column. These size-separation columns can be calibrated with a range of proteins of differing molecular weights and our target compound's molecular weight read from its retention time on the column. However, we often get an incorrect answer because gel permeation columns separate based on the Stokes radius of the protein, and that is greatly influenced by protein folding to form its secondary structure. We can get a more accurate molecular-weight number from denatured protein by running the column with 0.1% sodium dilauryl sulfate (SDS) in the mobile phase. The protein generally loses any enzymatic activity under these conditions, but we obtain a linear relationship between retention times and molecular weights of protein standards.

For structure determination we need a more accurate measure of molecular weight than we can obtain from a gel column. Light-scattering detectors can provide such an answer, but the most useful analyzer for this purpose is the mass spectrometer. Using an ESI-LC/MS system, we can generate a collection of molecular ions from a single protein (Figure 13.1).

Most proteins contain side chains containing basic amino acids such as lysine and histidine. These can support a positive charge like the N-terminal amino acid, so we can have multiply charged proteins. The mass spectrometer's analyzer separates on the basis of m/z (i.e., mass divided by charge), so our purified protein shows an entire family of molecular ions with different charges. Typically, a protein will exhibit molecular ions with a charge range of 7 to 22 charges. We can use a computer to measure the mass differences between peaks and

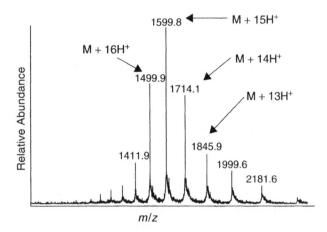

FIGURE 13.1   Protein molecular ion envelope. (Courtesy of BioAnalytical Systems.)

run deconvolution software to provide an accurate measurement of the original protein's molecular weight by averaging the molecular weights determined for each peak. These measurements are often made using a time-of-flight analyzer that can analyze very large molecular weights, but even quadrupole analyzers can be used for this purpose for enzyme molecular-weight determinations, due to the multiple charges involved. Because of multiple charges on the protein, the m/z value of individual peaks usually will fall within a mass range of the quadrupole of 0 to 3000 amu.

13.2   DE NOVO PROTEIN PURIFICATION

In addition to biological proteins appearing in nature, there are two other sources of protein that offer pharmaceutical possibilities. Proteins can be synthesized by the Merrifield synthesizer method, in which an amino acid is attached to a polymeric bead and peptides and proteins are built up by sequential condensation of other amino acids. Once synthesis is complete, the final protein is released from the bead and must be characterized and purified from any failures from incomplete condensation steps. LC/MS and LC/MS/MS are used to prove the nature of the final product and provide confirming molecular weights.

Expressing purified sections of DNA and c-DNA to identify the function of genes can also produce proteins. Once the 1,000,000-mer gene for muscular dystrophy was isolated, it was used to produce mRNA, which was then used to produce a protein called dystrophin. It was found to be identical to a protein present in muscle cells as the connector between the cell wall and the actinomyosin complex responsible for muscle contraction. LC/MS can provide molecular-weight and structural information after separating the protein from reagents, nucleotides, and restriction fragments.

13.3 PROTEIN ANALYSIS BY TWO-DIMENSIONAL
GEP AND LC/TOFMS

The protein of interest must be separated from the matrix in which it occurs, the protein fraction separated, and then the target protein broken into pieces for determination of its primary structure. The current most common method of separating proteins for derogative analysis is by using two-dimensional gel electrophoresis (GEP). The protein mixture is spotted on the bottom of a poly-acrylamide plate and run in a buffer solution under an electrical charge voltage differential to produce a first-dimension separation. The plate is then turned 90° and electrophoresis is run in the second dimension in a denaturing buffer, usually containing SDS. The plate is then sprayed with a visualizing spray for proteins to produce a two-dimensional array of the proteins present.

Known compounds are eliminated by comparison to standards and the spots representing target compounds are scrapped into tubes, extracted to remove the staining compound, and then treated with enzyme to cleave the protein into peptides. The most commonly used enzyme for this protein digestion is trypsin, although a number of other enzymes could be used. Trypsin attacks the peptide bond associated with basic amino acids and produces reasonably sized peptides.

The tryptic digest is separated in the liquid chromatograph, and the effluent is spotted on a plate array, mixed with a matrix containing a chromaphore such as 3-cyanocrotonoic acid, and dried. The plate is introduced into the ionization source chamber of a time-of-flight mass spectrometer and pulse-bombarded by a laser. The chromaphore absorbs the laser light and explodes into a source chamber, ionizing and carrying the sample into the analyzer's flight tube for separation and detection. This explosive mazing of the compound gave the technique its name, matrix-assisted laser desorption ionization (MALDI). This clumsy-appearing sequence has been automated using a robotic workstation for recovering proteins from two-dimensional EP gels, for tryptic digestion, for injection into the LC, for mixing matrix and effluent on the spotter plate, for evaporating, and for transferring the proteins to the MALDI/TOF source. Very fast laser response and TOF flight times allow multiple analyses from a single spot on the plate, providing very accurate analysis of the peptide mixtures that are produced (Figure 13.2).

13.4 LC/MS/MS IDENTIFICATION OF PEPTIDE STRUCTURES

Molecular-weight determination of separated peptides provides insufficient infor-mation to characterize their complex structure or identify them. Usually, this would require isolation and analysis of their amino acid sequence using Edman degradation studies. However, LC/MS/MS can provide structure information on the peptides by doing fragmentation studies and daughter-level scans. Fragmen-tation libraries of previously determined peptide structures have been digitized and collected, and can be computer searched for comparison and identification

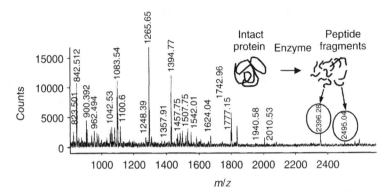

FIGURE 13.2 MALDI/TOF TIC of tryptic peptides. (Courtesy of BioAnalytical Systems.)

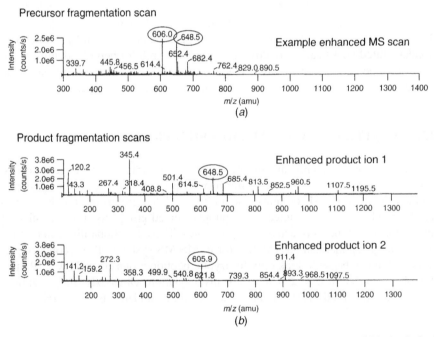

FIGURE 13.3 (a) Library ID and (b) peptide daughter ions. (Courtesy of BioAnalytical Systems.)

(Figure 13.3). This allows rapid identification of the tryptic digest peptides, and existing software routines can be used to look for overlapping areas in peptides and help to generate possible structures for the original target protein in the two-dimensional EPG spot.

13.5 TRACER LABELING FOR PEPTIDE ID

Mixtures of proteins and peptides separated by a LC/MS system are very complex, and the total ion chromatograms indicate much more complexity than we are able to analyze. A poster at a recent HPLC meeting shows work done on serum proteins using an LC/TOFMS/TOFMS and a unique computer algorithm capable of removing detected peaks for analyzing the next run. It found 284 protein peaks in the first run, 327 more peaks when these were omitted, 370 in the third pass, and 381 peaks on the fourth pass. A similar type of problem is often seen in examining peptide peaks from tryptic digests.

Separation tools have been designed for simplifying such mixtures. For example, one isotope-coded affinity tag reagent is designed to react with proteins or peptides containing cysteine groups. It contains a cysteine reactive group, an eight-deuterium hydrocarbon chain in its balance group, and a cleavable biotin reporter group. The peptide mixture reacts with equal amounts of this reagent and a nondeuterated equivalent reagent, the mixture is purified on an affinity column specific for biotin, the biotin functional is cleaved with acid hydrolysis, and the simplified peptide mixture containing both labels is analyzed by LC/MS. The chromatogram is made up of a series of peak pairs separated by 8 amu. The nondeuterated peaks can be referred to a library search for peptide identification. Other isotope-coded reagents are being developed and are starting to appear in the literature for attacking other specific side chains for simplifying peptide mixtures.

13.6 POSTTRANSLATIONAL MODIFIED PROTEIN

Not all proteins of interest are simple sequences of amino acids. In some cases, functionality is added in the cell after a protein is synthesized. These posttransitional modified (PTM) proteins contain glycosylated groups of complex sugars used for cell recognition indicators or phophorylated groups used inside cells for things like attachments or active transport pores. These functionalities can provide a specific group that can be searched for in MS/MS systems. The secondary precursor scan can be set to look only for tryptic peptide groups that show a 79 mass for phosphorylation or a 204 mass for glycosylation. Obviously, these peptides need to be identified, marked, and excluded from standard library searches.

13.7 TRANSIENT PEPTIDES AND ACCUMULATION PROTEINS

Peptide hormones are very important regulators of many of the body's functions. Peptides produced in the pituitary and hypothalamus in the brain and in the lungs and kidneys work through feedback mechanisms to control regulatory systems such as growth, appetite, blood pressure, and immune response. All share a common characteristic: They persist in the body for only a short time and are difficult to study on-line and in real time.

The second classes of proteins that provide a research challenge are fossil proteins: oxidized muscle proteins in heart and plaque proteins in the brains of Alzheimer patients. A study of the heart muscle of an autopsy cadaver showed that 70% of the protein in the heart muscle of a 70-year-old man is oxymyoprotein, nonfunctional oxidized muscle protein. Plaque protein formation is the current candidate for producing the symptoms seen in Alzheimer dementia. What happens to cause formation of these proteins, and how do you study them in a living human being?

LC/MS may offer some potential for providing answers in these two dissimilar areas of proteomics. Miniaturization, ultrafast chromatography, and fast-response, high-precision mass spectrometry such as provided by FTMS may offer ways of approaching studies of these problems in blood circulation, spinal fluid, or intact tissue biopsy.

14

THE FUTURE OF LC/MS

Liquid chromatograph/mass spectrometry is the ideal technique for separating mixtures in solution and providing a definitive identification of its components. Current limitations of the technique are primarily in separation speed, peak resolution, system size and cost, and in the inability to resolve and detect very large biomolecules. From the viewpoint of the organic biochemist and molecular biologist, this technique in a simplified form belongs on every laboratory bench. Accelerating growth of LC/MS system penetration of the laboratory equipment market will require systems to be simpler, cost less, take up less lab bench space, and be easier to tune and calibrate. Development of reproducible turnkey methods for quantitation and simplified library searching for MS/MS fragment library identification will help drive LC/MS into becoming a generalized laboratory and quality control instrument. But of most importance is system cost reduction.

14.1 INSTRUMENTATION IMPROVEMENTS

Current LC/MS systems spend much of their time waiting for the HPLC separation to be completed. This is especially true of advanced time-of-flight, Fourier transform, and linear ion trap analyzers. Improvements in LC column run time while maintaining or improving resolution will go a long way toward improving LC/MS use. Pharmaceutical development laboratories seem to be willing to sacrifice resolution to gain analysis speed, as do the clinical labs at St. Olaf Hospital (Chapter 10). But this trade-off may not always be acceptable in analyzing complex mixtures.

LC/MS: A Practical User's Guide, by Marvin C. McMaster
Copyright © 2005 John Wiley & Sons, Inc.

HPLC manufacturers are trying to improve separation times by moving to ultra-fast HPLC systems. These use very small-particle-size HPLC packed columns run at very high system pressures. Silica has always had the mechanical strength to handle pressures to about 15,000 psi. The problem has been in sealing the system to resist leakage. As column diameters moved from 10 μm to 5 μm to 3 μm, it was found that resolution loss began to level off at a high mobile-phase flow rate. Particles with diameters of 1 to 3 μm show very little, if any, resolution loss at very high flow rates. Of course, very small-diameter particles still show very high backpressures at high flow rates, require very small pore frits to keep them in the column, and are very easily contaminated with particulate matter in the mobile phase. Systems are now being sold that are optimized for LC/MS operation at 12,000 psi backpressure and flow rates of 6 to 10 mL/min, with a corresponding decrease in separation times.

Elevated temperature has been used to decrease separation dramatically for zirconium columns, since these columns are not dissolved on heating in aqueous solvents such as silica-based columns. Run times have been decreased three- to tenfold by using special system heaters that provide constant temperature control for injector line, column, and detector flow cells. These elevated temperature systems are being interfaced into LC/MS systems in pharmaceutical accounts.

Another time bottleneck in LC/MS analysis is the use of two-dimensional gel electrophoresis in the resolution of protein mixtures. Both electrophoretic dimensions must be run and the spots scraped and extracted before they can be reacted with trypsin to form peptides. A number of techniques are appearing in the literature for two-dimensional liquid chromatography using ion-exchange and reverse-phase columns to separate proteins with a high-enough resolution that they can be used for trypsin digests. These studies may also lead to LC/LC/MS/MS research systems for direct characterization of protein structure from fragmentation libraries.

One interesting development in LC/MS interface design appeared in 2000 in a paper by Amirav and Granot on supersonic molecular beams. The effluent stream from the HPLC is vaporized at atmospheric pressure at flows up to 0.25 mL/min and expelled through a supersonic nozzle into a vacuum system over a fly-through 70-eV electron ionization filament. The fragment ions formed can be separated in a quadrupole analyzer and used to search existing NIST libraries for compound identification. Initial work shows promise to produce a simple LC/MS that requires no interface nebulizer gas and produces precursor fragmentation scans for compound identification by library searching.

The success of this innovation is not too surprising. One of the first LC/MS systems was built by adding an ion spray interface to a mass spectrometer designed for GC/MS with a 70-eV filament source. It was found that turning the filament on and off during scanning and averaging the fragmenting and nonfragmenting scans produced a fragmentation pattern with a strong molecular ion. Like a normal ISI-LC/MS system, these scans provide the molecular weight of the compound, but this fragmentation pattern could also be used to search existing mass spectral libraries to identify compounds correctly, like a precursor scan generated by an LC/MS/MS system.

Systems like these are a poor person's LC/MS/MS, providing both molecular weights and precursor scans for compound identification. They cannot run all of the experiments that can be run by hybrid LC/MS/MS systems, but they would provide the information needed by the average laboratory chemist with the proper software. Many of these requirements are needed to make an LC/MS system a general laboratory analytical tool for the average user in solving research problems without becoming a research problem.

14.2 AFFORDABLE BENCHTOP LC/LITMS

As long as we are drawing up a laboratory wish list, I would like to propose the ideal next-generation benchtop LC/MS and LC/MS/MS systems. A generic LC/MS system for the mass market has to be easy to use, and the information it generates has to be relatively easy to interpret. It should be a tool for getting research or quality control information, not a research project in itself. It should be able to do autotuning and autocalibration procedures with standard compounds injected through the interface.

A low-end LC/MS system would use a four-solvent gradient LC system. Its standard ion spray interface could be quickly exchanged with an optional electrospray interface for proteins and ionized samples. The interface would feed a linear ion trap mass spectrometer capable of being run as either a scanning quadrupole or a trap for SIM operation or simple daughter-ion fragmentation. This small-footprint dedicated system would cost less than $100,000.

An LC/MS/MS system would use a fast-flow, high-pressure HPLC using a Peltier temperature-controlled spotter plate autosampler to protect proteins in solution and be capable of running hybrid silica, zirconium, and reverse-phase polymeric columns. The mass analyzer would be a short quadrupole interfaced through a collision cell into a linear ion trap capable of being run either as a linear ion trap (LIT) for molecular-weight determination, or as a QqQ MS/MS or QqLIT MS/MS/MS system for fragmentation studies.

The data/control modules for both systems should be able to display TICs or strongest-peak chromatograms and daughter scan fragment peaks all labeled with m/z. The more advanced systems should be able to display TICs overlaid with neutral loss or mass offset peaks and do library searching for precursor compound identification.

14.3 USER-CUSTOMIZED DATA LIBRARIES

A fragmentation library should be available on a real-time point-and-shoot search basis during display of a chromatogram. Standard libraries of mass spectral data such as NIST and Wiley contain information that the average user will never use. In an entire working career, this hypothetical chromatographer will probably be interested in only about 10% of the compounds in a library. What a user is interested in is the type of compound that he or she is working with at any given time and its impurities and metabolites. A customizable mass spectral database could be updated, like a word processor's spell checker's library, with the chromatographer's own fragmentation patterns run on his or her own LC/MS and with the compound structure identifications that the user has assigned. The longer the chromatographer uses it, the better the accuracy on the compounds of interest, assuming that the user's structure assignments are correct.

14.4   NUCLEOMICS AND RESTRICTION FRAGMENT ANALYSIS

New application areas are difficult to predict. Certainly, many of the applications areas mentioned in Chapter 10 will move to LC/MS as less expensive systems and certified methods become available. One area that will benefit from LC/MS/MS for sequencing and structural determination is the study of DNA, RNA, and posttranslational modified nucleotides such as those found in cancer cells. Restriction fragments are already being analyzed by LC/TOF-MS, but the ones that can be separated and detected are limited in size. Polynucleic acids in solution are long, straight rods with large masses. Each nucleic acid added increases the mass by 250 to 300 amu versus about 100 amu for an amino acid added to a protein chain. A recent paper by Chalk and Matter indicated that a useful limit of 30 kDa or a 100-mer is the largest nucleic acid that can routinely be separated and detected. Detector technology is the main reason for the limit, due to the nature of the electron multiplier detectors used in mass spectrometers. High-mass-biopolymer ions reach the detector with lower velocity, reducing the detector response. Polynucleotides show a 1000-fold lower response than that of peptides. The authors were able to overcome this with a commercially available cryostat superconducting detector that showed 100% response on large biopolymers and detected protein complexes with up to a 2,000,000-Da mass.

APPENDIX A

LC/MS FREQUENTLY ASKED QUESTIONS

Listed in this appendix are common LC/MS questions from students and customers and my answers. This list is not exhaustive. One of the most common questions that I did not include was: "Why won't my system start up?" I would ask the person on the phone if the system was plugged in. After the explosion on the other end settled down, I would say: "Sir, sometimes janitors unplug lines so they can plug in their polishers. Would you please check to see if it is plugged in?" Often, after about a minute or so, I would hear a quiet, embarrassed click as the phone was hung up.

A.1 HPLC FAQs

1. *Why do I need to use helium gas on a liquid chromatograph?* Helium might be used for two reasons. Low-pressure-mixing-valve gradient systems suffer from bubbles being pulled out of solution and stalling the pump head unless air is flushed out of the solvents by helium purging. Sometimes the solvent reservoirs are pressurized with helium gas to aid in smooth solvent flow. Helium or nitrogen may also be used as the nebulizer gas in an atmospheric-pressure ionization interface to remove solvent, volatile buffers, and aid in ionizing compounds in the LC effluent.

2. *Do I need a gradient system, and if so, why?* Gradient systems let you control flow rate and solvent/buffer changes to improve chromatographic separations. They can be used to sharpen separations and to speed column reequilibration. A

LC/MS: A Practical User's Guide, by Marvin C. McMaster
Copyright © 2005 John Wiley & Sons, Inc.

four-solvent gradient system is useful for methods development when equipped
with methanol, acetonitrile, ammonium acetate buffer, and formic acid solution.
But many quality control laboratories prefer to use inexpensive isocratic systems
that run a constant-composition premixed mobile phase for rapid separations.

3. *Do I need an autosampler?* Autosamplers and robotic workstations provide
reproducible injections and allow automation of the chromatographic separation,
but add significant cost to a system. Many university laboratories prefer to substitute graduate students to do the job.

4. *Why does my LC system keep shutting itself off?* HPLC pumps are equipped
with an overpressure setting to protect fragile columns. Perhaps your settings are
too low, or your column frits may be plugged, providing too much backpressure.
If the pressure settings are correct, you may have to clean the line from the
injector to the column or the column frit. The most common cause of plugs
in lines or frits is material not filtered from samples or mobile phase. Buffers
precipitate when you switch between incompatible solvents. Washing buffers out
with water before moving to a new organic mobile phase will help prevent this
problem in the future.

5. *What kind of sample preparation do I need to do before I inject?* That
depends on what you are analyzing. As much of the interfering compound(s)
needs to be removed as possible: proteins precipitated, lipids extracted, cells
and particulates filtered or removed. Some samples need to be concentrated to
aid in detecting trace amounts in dilute samples. Check the literature for your
particular compound, use traditional procedures for compound purification, and
look into the possibility of using SPE columns for precolumn purification and
concentration.

6. *Do I need HPLC-grade solvents and water? Do I need to filter samples?*
The answer to both of these is "yes"! HPLC water is the most important ingre-
dient. Triple-distilled water used for HPLC has ruined many a chromatography
run because of co-distilling of nonpolar organic contaminates. Use HPLC-grade
solvents and columns from a reliable supplier. Unfiltered samples lead to injector and column frit plugging. Filtering mobile phase after premixing will cut
wear and downtime on pump check valves and lines. If you are going to invest
money in systems and operators, do not cheapen the system by trying to save
money on supplies. Also, pick a column manufacturer and stay with that firm
to get the best reproducibility from column to column. Otherwise, you will continuously be wasting your lab's time doing methods development to adjust your
chromatography to the new column.

7. *Which is the best column: reverse phase, ion exchange, or size separation?*
Each column type has its own use. Column variety is what gives HPLC its versatility. It really depends on your compound and application. Approximately 80%
of all separations are done on a 5- to 10-μm reverse-phase C18–silica column.
Much of this is tradition. A reverse-phase column offers high-resolution separations for a wide variety of compounds and can be run in aqueous mobile phases.
Ion-exchange separations require salt solutions for separations, and these are not

popular in mass spectrometers. Size separations have lower resolving power and longer run times but may be the only way to separate proteins solutions that will irreversibly stick to reverse-phase columns. Use small-pore size separation columns to remove salt from effluent from other chromatography separations. Zirconium and polymeric columns are newer and offer possibilities for unique separations.

8. *Do I have to use buffers to achieve separation? What is ion pairing?* Buffer and ion-pairing reagents are used to sharpen and control separations. Buffers adjust the pH of the mobile phase and the compounds it contains. Compounds such as organic acids, phenol, and amines are partially ionized at their pK_a value. They exist as two species that the column tries to separate, leading to broad, tailing peaks. By using buffer to control pH, we force the compound into one form or the other. For LC/MS, choose a volatile buffer such as ammonium acetate instead of a nonvolatile such as sodium phosphate. Ion-pairing reagents form nonbonded complexes with ionized compounds to control where they elute in a separation. If possible, avoid them in LC/MS, but if you must use them, use volatile forms.

9. *How long will my HPLC columns last?* An HPLC column has no specific life span. A number of things can "kill" columns, including bonded-phase loss, voiding or channeling, dissolving column packing, and irreversible binding of material from injections. Most can be either prevented or treated. Many "cost-per-test" commercial laboratories set a goal of 1000 injections for a column. Silica columns should be kept as much as possible at a pH between 2.5 and 7.5, and pressure shock and sudden changes of solvents should be avoided. Zirconium column are more forgiving of pH extremes. Polymeric columns can be used from pH 1 to 11 but should be protected from high-pressure bed collapse. For more information on column killers and column healing, see my book, *HPLC: A Practical User's Guide.*

10. *How long does it take to change columns?* Column changes are a feasible method development tool in HPLC. A column can be switched at any time. Attach the new column with a slow flow of solvent from the injector outlet line to fill the inlet fitting of your column so that you do not leave air in the column entrance. Figure on flushing a column with at least six columns of a new solvent to reequilibrate the column before injecting a sample. Plan on ignoring the first injection. Usually, the second, third, and subsequent injections will be reproducible.

11. *Why does my new column give such bad-looking peaks?* You either used the wrong diameter of tubing for the line from the injector to the column or you used tubing from a previous column connection. Manufacturers use different depths in column fittings. Tubing of the wrong diameter or a line made for a different column can create extra column volumes that can ruin your chromatography. Use fine-internal-diameter tubing and always make the fitting in the column inlet in which it will be used: 0.009-in. tubing for 10-μm packing and smaller-diameter tubing for 5- and 3-μm packing. I once saw a brand-new 5-μm column give terribly broad tailing peaks in a laboratory. The inlet line had come

from a mixed tubing drawer and I found it to be 0.02 in. ID. The column gave perfect needle-sharp peaks when we replaced the line with 0.009-in. tubing and a fitting made in the inlet hole.

12. *How do I go about cleaning columns?* Column cleaning is an art covered extensively in *HPLC: A Practical User's Guide* and in some detail in Chapter 4 of this book. Most things that adhere to a column will stick on a guard column of the same packing. Guard columns are cheaper to replace than analytical columns. Remove bound material from a C18 partition column by washing with the solvent that is most like the column in polarity. First wash the column out with water if you a using a buffer in the mobile phase before doing the organic washout.

13. *My chromatography has changed. How can I fix it?* Chromatography changes because the column is either degrading or something is sticking on the column and needs to be washed out. Controlling pH and pressure shock on a silica column will help prevent degradation. Washing with solvents of the same polarity as the column will remove bound organics. High-concentration salt washout will remove most things from ion-exchange columns.

14. *What is pacification, and how can it help protect my LC?* Pacification is a technique for removing organics and buffers from HPLC metal and Teflon surfaces and protecting them from salt corrosion with 6 *N* nitric acid (see *HPLC: A Practical User's Guide* and Chapter 4 of this book). **First, remove the HPLC column and replace it with a column bridge. Do not flush this wash into the mass spectrometer.** Wash the system with water. Remove the column and replace it with a column bridge. Flush with 6 *N* nitric acid for at least 30 minutes, then overnight with water. Ensure that the effluent pH is back to the pH of the lab water. Replace the column and flush with mobile phase. This should be done at least once a month to clean check valves, line, and injectors. **Under no circumstances should this wash be done with an HPLC column in place or into the mass spectrometer!**

A.2 INTERFACE FAQs

1. *Which is better: ion spray, electrospray, or nanospray?* Each interface has its use in LC/MS. Ion spray works best with neutral and nonpolar compounds. Electrospray is preferable with ionized or polar materials such as proteins. Nanospray is electrospray optimized to give the best production of ions for polar compounds from microflow columns, especially when only trace amounts of sample are available.

2. *What is that liquid running off on the floor?* The MS interface is designed to ionize only a tiny portion of the available sample. Part of the LC mobile phase is vaporized, but much of it is left behind and can be diverted to a secondary detector. At 1 to 2 mL/min, this represents a large volume of solvent that must be diverted to some type of reservoir for disposal or mopped up off the floor.

3. *What secondary detector should I choose?* Most laboratories use a variable UV detector or whatever is left over from other HPLC systems. Chromaphores

absorb at 254 nm, and almost any compound shows some absorption in the UV spectra at 195 nm, but you have to use acetonitrile and phosphate buffers to operate at a wavelength that low. Volatile buffers in the mobile phase preferred for use in mass spectrometer interfaces prevent operation much below 225 nm. ESA has just released a universal detector called a corona charged aerosol detector with excellent sensitivity ideal for gradient work with volatile buffers that should be a good complement to mass spectrometer operation. It allows detection of compounds such as sugars, phospholipids, and steroids that are very difficult to see with UV.

A.3   MS FAQs

1. *Which mass spectrometer is best: quadrupole, ion trap, or time of flight?* Each analyzer has its own strengths. Quadrupoles were the first mass spectrometers used for chromatography, still represent the bulk of laboratory systems, and are the least expensive. Look for a benchtop quadrupole that can give you chromatograms labeled with peak molecular weights. Linear ion trap analyzers are the most flexible, allowing you to determine simple molecular weights as well as precursor and product fragmentation studies. They can also store enough ions to give excellent sensitivity for trace analysis and can trap specific ions for fragmentation analysis. For large-molecule studies, you are going to want a time-of-flight system with an electrospray interface.

2. *How high a vacuum should I be getting?* You will need at least 10^{-5} torr before you begin running chromatography and should see 10^{-6} torr on a clean system with a turbo pump. I have talked to people who get 10^{-7} torr with an oil diffusion pump and clean oil. Sometimes you need to push down on the lid on the mass spectrometer containment system if you are having trouble making a vacuum after cleaning the system. Try changing the pump oil in your rotary vane pump if you are not able to reach 10^{-5} torr with the turbo pump.

3. *How do I service my system's turbo pump?* A turbo pump is a miniature jet engine and should be worked on only by a trained technician. The best plan is to pay for an exchange program warrantee with the system manufacturer. When a turbo pump is down, it is down. It is nothing you want to mess with repairing.

4. *How often do I need to replace my ion detector?* Cascade-type detectors have a finite lifetime and need to be replaced periodically. Record the EM voltage needed to tune a new detector and watch that value as you continue to autotune. When it exceeds 3000 V, look for a replacement. Photoarray detectors used on TOF systems have fewer problems but are more expensive. Problems show up as increased background noise, due to increasing element chatter.

5. *How do I know when I need to clean my mass spectrometer?* Dirty systems lose sensitivity and become very difficult to autotune. There are fewer problems with analyzer contamination in an LC/MS system than a GC/MS, where most of the problem is char due to filament-based sample burn. However, both types of system accumulate organic contamination on the analyzer and detector

surfaces. Some labs disassemble the system, immerse the analyzer rods (still in their ceramic collars) in a graduated cylinder, and flood them with methanol. The detector face is then wiped with lint-free paper wetted with methanol. The analyzer rod package is dried in an evacuated vacuum desiccator if the rods are small enough. If not, they are wiped with lint-free paper and dried by pumping overnight in the mass spectrometer containment chamber.

6. *All I want is molecular-weight chromatograms? Will my peaks be labeled?* The peaks will be labeled with molecular weights if you select the correct software. APCI interfaces such as ion spray and electrospray are low-energy ionization sources and generally do not fragment molecular ions. Some software will select the major fragment ion and label the TIC with major ion molecular weights if running an ion trap detector to produce fragments or running an MS/MS system.

7. *How come I don't see ion fragments in my chromatographs?* As I mentioned in question 6, APCI interfaces product low-energy ions and usually do not fragment the molecular ion unless you are running some kind of high-voltage filament or a repeller plate. Fragmentation occurs in GC/MS because it uses a 70-eV filament voltage to ionize the sample, and that energy produces fragments from the molecular ion breakdown. Fragmentation occurs in MS/MS by collisional energy from ions striking large gas molecules.

8. *I do protein separations. Do I need MALDI/TOF or LC/MS/MS?* Probably you will need access to both types of systems. See if you have a shared LC/MS facility available to you to aid your research. MALDI/TOF/MS systems are used to analyze proteins and trypsinized peptides from gel electrophoresis scrapings. You may need to do precursor and product scans to confirm peptide identities, especially if you are working with PTM proteins.

A.4   LC/MS FAQs

1. *How do I start up my system after it has been sitting unused for awhile?* Check the operator's manual that came with the system. I would start up a system I am unfamiliar with by using the following procedure. It really depends on whether the vacuum system on the mass spectrometer has been turned off. If it has, you will need to start the rotary vane oil pump and pump until you reach a vacuum of 10^{-4} torr. Now turn on the turbo pump and begin to establish your analyzer vacuum. This may take a number of hours.

Check to see if the HPLC column is in place. Hopefully, someone has removed the column and capped it if the system has been inactive for awhile. Place a beaker under the injector outlet and use 25% methanol to prime the LC pump with the injector or autosampler in the load position. Start the pump flow slowly if the system has sat for awhile to slowly wet the Teflon surfaces and the check valves. Degas and filter your mobile phase(s) and switch out your priming solvent. Remove the column caps, and with solvent running, connect the injector line to the column inlet. Next, connect the column to the interface module, making sure that you have the overflow outlet in a reservoir. Select your flow rate and start mobile-phase composition for your chromatography run.

Check to see if your vacuum has reached at least 10^{-5} torr. Turn on the nebulizer's gas, inject your calibration standard into the interface, and run an autotune on the mass spectrometer. By now your column should be equilibrated and you are ready to make your first injection. Set your mass range selection for scanning and start your autosampler, solvent gradient run, and data acquisition from the control computer.

2. *Can I fix up my GC/MS to do LC/MS?* The first LC/MS systems were made from a mass spectrometer converted from a GC/MS, so you should have little problem. You will need an HPLC system and an ionization interface, so this will not be a trivial conversion. It really depends on how much you use the GC/MS system. Switching back and forth will rapidly become a pain. At a minimum you will have to shut off your vacuum, remove the GC column, insert the ionizing interface probe, and then reestablish the mass spectrometer vacuum. You might want to consider writing another grant proposal and buying a dedicated system after you have made the conversion one or two times.

3. *What is the difference between calibration and autotuning?* Calibration involves adjusting the mass axis to align it with known peak positions from an injected standard using tuning lens, amu offset lens, and the electron voltage of the detector. Tuning involves balancing the settings on the same lens to adjust the relative heights of the calibration peaks to a predetermined relationship. Auto-tuning is done by the system controller adjusting the lens, and other components to produce a specific calibration and tune when calibration standard is injected through the interface port or from the LC system. The operator's manual for your particular system should recommend calibration standards and tuning criteria. If it is not in the manual, contact their technical support staff or their Web site. Your instrument's calibration and tuning need to be checked periodically if your MS results are to be compatible with those from other instruments and useful for fragmentation database library searches.

4. *When do I need LC/MS/MS, and when will LC/MS be good enough?* An LC/MS system will provide you with chromatograms of column peaks and can provide molecular weights of the separated materials and their solvent complexes. An LC/MS/MS system will be much more expensive but can provide structural information to confirm the identity of the compounds separated. Unfortunately, it will do it at a cost. Someone will have to plan the fragmentation study and interpret the results. A library database may be able to make this identification for you for simple compounds, but generally, each sample run will be something of a research project.

5. *Can I use my NIST library to identify compounds in my chromatogram?* An LC/MS system generally does not fragment the sample in the effluent and usually provides only a single ion equivalent to the molecular weight unless a solvent conjugation artifact is present. You will have to use an ion trap LC/MS system or an LC/MS/MS to get fragmentation information that can be searched by your

NIST library. It is important to remember that the NIST and other commercially available libraries were originally developed from runs made on magnetic sector and quadrupole GC/MS systems. LC/MS data are being added, so make sure that you are running an up-to-date library. LC/MS manufacturers try to adjust their fragmentation analyzers so that they are compatible with existing libraries, but there will be differences. Hopefully, these will not prevent you from using your library.

APPENDIX B

SOLVENTS AND VOLATILE BUFFERS FOR LC/MS

B.1 LC/MS SOLVENTS

A list of LC/MS solvents is given in Table B.1. Note that some solvents are immiscible. Polar solvents exhibit viscosity changes in aqueous gradients that can cause major pressure effects during a run and make trigger overpressure settings.

B.2 LC/MS VOLATILE BUFFERS

A list of LC/MC volatile buffers is given in Table B.2. Note that 1 to 10 mM buffer concentration is usually recommended for LC/MS.

TFA is known to quench ionization in electrospray LC/MS, leading to lower sensitivity, and should be avoided.

B.3 SOURCES OF LC/MS BACKGROUND CONTAMINATION

The LC/MS is an extremely sensitive instrument. However, achievement of this kind of sensitivity is background dependent and requires elimination of all common sources of contamination.

LC/MS: A Practical User's Guide, by Marvin C. McMaster
Copyright © 2005 John Wiley & Sons, Inc.

TABLE B.1   LC/MS Solvents

Solvent	Formula	MW (Da)	Boiling Point (°C)	UV Cutoff (nm)
Acetonitrile	CH_3CN	41.05	81.6	190
Chloroform	$CHCl_3$	119.38	61.7	245
Dichloromethane	CH_2Cl_2	84.93	40.0	235
Ethanol	CH_3CH_2OH	46.08	78.5	210
Ethyl acetate	$CH_3CO_2CH_2CH_3$	88.12	77.1	260
Diethyl ether	$(CH_3CH_2)_2O$	74.12	34.5	220
Heptane	$CH_3(CH_2)_5CH_3$	100.21	98.4	200
Hexane	$CH_3(CH_2)_4CH_3$	86.18	69	200
Isopropanol	$CH_3CH(OH)CH_3$	60.11	82.4	210
Methanol	CH_3OH	32.04	65	205
n-Propanol	$CH_3CH_2CH_2OH$	60.11	97.4	210
Tetrahydrofuran	C_4H_8O	72.12	66	215
Toluene	$C_6H_5(CH_3)$	92.15	110.6	285
Water	H_2O	18.02	100	none

TABLE B.2   LC/MS Volatile Buffers

Volatile Buffer	Structure	pK_a	Buffer Range
Trifluroacetic acid	CF_3CO_2H	0.5	3.8–5.8
Formic acid	HCO_2H	3.8	—
Ammonium formate	HCO_2NH_4	3.8	2.8–4.8
Acetic acid	CH_3CO_2H	4.8	—
Ammonium acetate	$CH_3CO_2NH_4$	4.8	3.8–5.8
4-Methylmorpholine	$OC_4H_8N(CH_3)$	8.4	7.4–9.4
Ammonium bicarbonate	NH_4CO_3H	6.3/9.2/10.3	6.8–11.3
Ammonium acetate	$CH_3CO_2NH_4$	9.2	8.2–10.2
Ammonium formate	HCO_2NH_4	9.2	8.2–10.2
1-Methylpiperidine	$C_5H_{10}N(CH_3)$	10.1	10.0–12.0
Triethylammonium acetate	$CH_3CO_2NH(CH_3)_3$	11.0	10.0–12.0
Pyrrolidine	C_4H_8NH	11.3	10.3–12.3

Essentially, there are two kinds of background that can interfere with trace-level LC/MS analyses:

1. *General background*, such as column bleed and hydrocarbons, which will generate a large TIC signal during the analytical scan and decrease the sensitivity level for detecting target compounds.

2. *Specific ions in the background* that interfere with a single ion or extracted ion chromatogram. For example, significant 164 background might be present when trying to detect low levels of 2,4-dichlorophenol. This type of problem is less common than general background contamination. Typically,

a single ion or an extracted ion can be chosen which does not appear in this background.

The easiest way to determine if the background is permanent is to run mobile phase from the HPLC into the interface and run a scan to see if the background decreases. If it does, the background is probably due to column bleed, contaminated pump oil, or leaks of various kinds.

In all instances where the background is determined to be coming from the analyzer and not eluting from the LC, the system should be shut down and the source cleaned. If this does not eliminate the problem, shut down the system and dip the rods, washing with methanol or methylene chloride to remove contaminants. A permanent background is defined as background that is at approximately the same level regardless of the material being pumped into the interface.

The other source of problems is contaminated samples (Table B.3). They tend to give more discrete chromatographic peaks and specific mass fragments. I usually try to remove these samples with SFE or SPE cartridge or GPC columns.

GPC columns separate on size and release smaller molecules before the larger, polymeric material. They are very good for removing "road tar–like" materials from your extracted samples, although getting the road tar off the column may prove to be a problem. Generally, if you can dissolve it, you can elute it.

The SFE or SPE cartridge columns are true chromatography columns. You can use them to do class separations of materials. Using windowing techniques and standards, you can work out methods for purifying the materials of interest from more polar or nonpolar contaminate. This technique is described in *HPLC: A Practical User's Guide*. The really nice thing about them is the cost. They

TABLE B.3   Common Contaminant Mass Ions

Mass Ions	Compounds	Source of Origin
18, 28, 32, 44	H_2O, N_2, O_2, CO_2	Air leak
28, 44	CO, CO_2	Hydrocarbon fragments
31	Methanol	Lens-cleaning solvent
43, 58	Acetone	Cleaning solvent
69	Fore pump fluid	Saturated trap pellets
69, 131, 219, 254, 414, 502	FC43 (PFTBA)	Calibration gas leak
73, 207, 281, 327	Polysiloxanes	Column bleed
77	Benzene or xylene	Cleaning solvent
77, 94, 115, 141, 168, 170, 262, 354, 446	Diffusion pump oil	Improper shutdown of pump heater
91, 92	Toluene or xylene	Cleaning solvent
105, 106	Xylene	Cleaning solvent
151, 153	Trichloroethane	Cleaning solvent
149	Plasticizer (phthalates)	Vacuum seal damage
14-amu spaced peaks	Hydrocarbons	Saturated trap pellets, fingerprints, pump fluid

are cheap enough to discard if you contaminate them and cannot wash out the contaminant.

Also check Table B.1 for solvent molecular weights and see if these may be contaminating your spectra. If you see solvent or buffer peaks when you are scanning at low m/z, it may indicate that your interface is not volatilizing them. Check to see if your gradient conditions are interfering with removal of volatile components. Consider moving to isocratic or dial-a-mix conditions to avoid dramatic effluent solvent change.

APPENDIX C

GUIDE TO STRUCTURE INTERPRETATION

I make no claims to being an expert in structural interpretation. This technique is a science in itself and beyond the scope of this book. I am presenting the information in this appendix as a guide to the way I have used this technique in unraveling some questionable library assignments.

Interpretation of molecular structures from an ion trap or from MS/MS fragmentation data is an involved, time-consuming, and exacting science. If you are the type of person who enjoys doing *New York Times* crossword puzzles, you might find it worth pursuing in more detail. An excellent book by McLafferty and Frantisek is designed to help you learn to extract structural information using ion fragmentation mechanisms. Table C.1 summarizes the main points and the order in which you need to acquire information from the spectra.

When you are running an LC/MS/MS system in a laboratory, how often do you need or want to do a rigorous interpretation of a structure? The answer is—almost never! It is faster, and a better use of your time, to do a library database search if one is available for your types of compounds and let the computer match your fragmentation pattern to known spectra.

This does not mean that structural interpretation is totally useless. Compound structures are important information. One of the problems with spectral library databases is that some of their structures are inaccurate or just plain wrong. The original interpretation of their structures may have been incorrect, or mistakes may have been made in entering them. The earlier Wiley database of 225,000 compounds was thought to have up to 8% incorrect structures in it. It is claimed

LC/MS: A Practical User's Guide, by Marvin C. McMaster
Copyright © 2005 John Wiley & Sons, Inc.

TABLE C.1   Guide to Molecular Structure by Fragment Analysis

1. Base peaks and relative ion intensities:
   a. Determine the molecular ion mass. Run CI if needed.
   b. A scarcity of major low even-mass ions = an even-mass MW.
2. Elemental composition from isotopic abundances:
   a. Look for A + 2 pattern elements (Cl, Br, S, Si, O). (Check the A + 1 ratios for the absence or presence of S and Si.)
   b. Use the nitrogen rule to determine the number of nitrogen molecules. (If the MW is even = 0 or an even number of nitrogen molecules; if the MW is odd = odd number of nitrogen molecules.)
   c. Number of carbon/nitrogen molecules from A + 1 isotopic ratios.
   d. Estimate the number of H, F, I, and P from A isotopic ratios and MW balance. (Only P is multivalent; F = 19 and I = 127 mass units.)
   e. Check the allowance for rings and double bonds. [The number of double bonds or rings $= x - \frac{1}{2}y + \frac{1}{2}z + 1$, where $(C,Si)_x(H,F,Br,Cl)_y(N,P)_z(O,S)$.]
3. Use molecular ion fragmentation mechanisms.
   a. Check fragment mass differences for expected losses (35 = Cl, 79 = Br, 15 = Me, 29 = Et, etc.)
   b. Look for expected substructures.
   c. Look for stable neutral loss (CH_2=CHR).
   d. Look for products of known rearrangements.
4. Postulate structures.
   a. Search library database.
   b. Run the hit compound on the same instrument to confirm.
5. Use MS/MS if further confirmation is needed.

---

that the current Wiley database have been cleaned up and that the structure assignments are >98% accurate.

Even when you are working with accurate known spectra and precise spectral matching algorithms, there are still sources of problems:

1. The tuning conditions used in preparing the target spectra may not have been the same as those you set up.
2. The spectra could have been run on an entirely different type of mass spectrometer with a different mass linearity. Some data in these libraries were run on magnetic sector instruments rather than on a quadrupole. The high mass areas of these two types of instruments do not calibrate in the same ways.
3. Either your spectra or the target spectra may have been run on impure compounds, which may introduce additional fragment peaks, especially at low relative intensities, which may affect matching.

4. You may have chosen to scan above 40 amu to avoid water and air peaks, while the target spectra may include these extra fragments altering the match.

The library matches do not give you a single compound; they provide you with a list of matching compounds similar to your spectra with a confidence level for the particular match. You may have a pair or more of possible structures among which you will be required to choose. Partial structure interpretation can be a useful guide to making a choice between close matches or in determining whether a high-probability match makes any sense at all.

C.1 HISTORY OF THE SAMPLE

The starting point for examining a fragmentation spectrum is to find out as much as you can about the compound you are examining: its source, its solubility, and whether its UV spectrum shows conjugation or aromatic structures. The chromatography can tell you a lot about its solubility, polarity, and hydrogen bonding when run under conditions that break such bonds. Look at the kinds of compounds with which it separates. Look up or measure its boiling point or melting point and determine the molecular weight from the LC/MS run. The more you know about the compound, the quicker you will be able to confirm the apparent match from the database.

Once we have its molecular weight and an idea of its chemical nature, we can move on to determine its elemental composition from isotopic abundance information calculated from fragment patterns. Finally, we examine the mass differences between fragments to determine what types of groups are being lost. If we do at least this much, we almost always have enough information to confirm a library structure assignment.

Molecular-weight information is available from the precursor compound's spectra. If the molecular ion is missing from the fragmentation pattern, being so unstable that it contributes at best only a very tiny peak, we can usually get it from the original API chromatograph. That will give us a molecular ion, and therefore the compound's molecular weight, as the first piece of information we must have to begin our analysis. The fragmentation pattern will point us toward whether to expect an even or an odd mass molecular weight. If you look through the fragmentation patterns and see a scarcity of major, even-mass ions in the low-mass range, you probably have an even-mass molecular weight. Be aware of the possibility of adduct molecular-weight information throwing off the molecular-weight determination. If you were running with ammonium ion buffers, you may get an $M \cdot NH_3^+$ adduct that may be giving a molecular weight that is 17 amu too high.

C.2 ELEMENTAL COMPOSITION

The next thing to determine is how many carbon, hydrogen, nitrogen, oxygen, and other elements are present. We can do this by looking for elements that show characteristic isotopic patterns in the fragment spectrum. You try to work with the most massive fragments and with the fragments having the tallest mass peaks. In any group, start with the most intense fragment in the group, the one with the most stable isotopes, as your A peak.

Table C.2 is a list of isotopic ratios for common elements making up organic molecules. A fragment containing an A-type element shows only a single band in the spectra. An A + 1 element, such as carbon or nitrogen, has two isotopic forms separated by 1 amu and forms a pair of fragment ions. The relative peak intensities of the fragments will be the same as the isotopic abundance of the element. If an ion fragment has a single carbon, the relative height of the first mass peak would be 100; 1 amu up would be a fragment with a height of 1.1. The effect is additive. The more carbon atoms in the ion fragment, the higher will be the second peak height (e.g., if five carbon atoms are present, the second peak would have a height of 5 to 6% of the first peak). If you do nothing else, find these A + 1 fragment pairs and use them to estimate the carbons present in the fragment. When you are working with organic molecules, you will be right much of the time. Biological molecules include enough nitrogen molecules to throw this number off.

Before we can work on carbon, nitrogen, and hydrogen, we must first determine the presence or absence of other elements. Fragments containing the A + 2 elements show a large peak, and 2 amu higher, a smaller peak of a precise height dependent on the quantity of that element in your compound. Chlorine stands out like a sore thumb. A primary fragment peak containing a single chlorine show a +2 secondary fragment that is one-third the height of the primary. Each fragment containing only one chlorine molecule will show this same A + 2 fragment ratio. This patterns occur because chloride is a mixture of isotopes, its major isotope has mass 35, but it has a second major isotope with mass 37 with 32% of the 35 mass isotopic abundance. You can tell when a fragment decays with loss of

TABLE C.2   Natural Isotopic Abundance

Element Type	Element	A Mass	A %	A + 1 Mass	A + 1 %	A + 2 Mass	A + 2 %
A	H	1	100	2	0.015		
	F	19	100				
	P	31	100				
	I	127	100				
A + 1	C	12	100	13	1.1		
	N	14	100	15	0.37		
A + 2	O	16	100	17	0.04	18	0.2
	Si	28	100	29	5.1	30	3.4
	S	32	100	33	0.79	34	4.4
	Cl	35	100	—	—	37	32.0
	Br	127	100	—	—	81	97.3

this chlorine. The mass difference between fragments will be 35 and the A + 2 pattern will not appear in the smaller fragment. Bromine shows an A + 2 doublet of almost equal height (100% and 97%). Compounds with multiple chlorine molecules or a mix of chlorines and bromines in the same molecule show other characteristic patterns that are the additive results of combining A + 2 patterns. Tables of these are available in Watson's book.

Once we have determined the number of chlorines and bromines present and have subtracted their contributions to the molecular weight, we need to look for the presence of S and Si. These are also A + 2 elements, but they show an additional isotope at the +1 position. First you find A + 2 patterns, then look at the mass in the middle. If there is no intermediate peak, you can scratch off these two elements. If there is an intermediate +1 peak, compare its height to the A + 2 peak height after you have removed any chlorine and bromine contribution, and compare it to the values in Table C.2. This should lead you to the number of sulfur, or less likely, the number of silicon molecules present. Oxygen is also an A + 2 element, but the isotopic contribution from ^{18}C is too low to be useful for measuring the amount of oxygen present in a fragment. Usually, it is estimated from the residual molecular mass after all elements except hydrogen are eliminated.

Once we have eliminated the A + 1 contributions from sulfur and silicon, we are ready to calculate the carbon and nitrogen values. In a simple hydrocarbon such as hexane, we should expect the +1 fragment peak to be about 6.6% as high as the main peak. Contributions by nitrogen are estimated using the *nitrogen rule*, which states that if the molecular weight is even, you will see either no nitrogen or an even number of nitrogen molecules in the fragment. Odd molecular weights occur when there is an odd number of nitrogen molecules. This allows us either to eliminate nitrogen or to come up with a satisfactory number of nitrogen molecules. Subtracting the nitrogen contribution should provide us with a good ratio of carbon isotopes, allowing us to calculate the number of carbons present. We can now subtract the carbon and nitrogen contributions to the molecular weight.

We are now left with H, F, I, P, and of course, oxygen. Because of its large isotopic mass, the presence or absence of iodine is usually obvious at this point and can usually be eliminated. Phosphorus is multivalent and most commonly bound to multiple oxygens, so it is usually easy to eliminate or identify phosphorus from this consideration. Fluorine's odd mass of 19 and its univalent replacement of hydrogen makes its presence or absence apparent when you are trying to distribute the residual molecular-weight units between oxygen, hydrogen, and fluorine. Once you have an elemental assignment in hand or even a partial that makes sense, check it against the compound selected by the library search engine. Does it agree?

One more check that can be done is to check the number of double bonds and rings that are present. Table C.1 presents a formula for calculating this number. You add up the number of quadravalent, trivalent, divalent, and monovalent atoms present and plug them into the formula. You end up with a number representing the total number of double bonds and rings present using the lowest valence state for the elements. For a benzene ring this number would be 4, for an electron-balanced charged ion, this number might be $\frac{1}{2}$.

C.3   SEARCH FOR LOGICAL FRAGMENTATION INTERVALS

The final things I look for in spectra are mass losses between major fragment peaks. I look for characteristic losses such as 35 (Cl^-), 15 (CH_3^-), 29 ($CH_3CH_2^-$), or a loss of 15 followed by a series of 14 ($-CH_2^-$), which indicates a breakdown of a straight-chain hydrocarbon. I look for neutral molecule losses such as substituted vinyl ($RCH=CH_2$) that occur during rearrangements; 79, which may indicate the presence of a phosphate group; and 28 (carbon monoxide), which may indicate the presence of a carboxylic acid or an aldehyde.

Once you find these markers, go back to the library structures and see if you can tell where these pieces are coming from. If none of these breakdowns make any sense, you may not have the right structure. If you can see how the pieces you are seeing can be formed, you have found additional confirmation for the structural assignment.

I hope this makes sense and helps you in confirming assigned structures. A rigorous study of fragmentation mechanisms will let you recognize many more loss assignments, but you will have to determine whether it is worth your time. In any case, the ultimate test is to acquire a sample of what you believe to be the correct compound. Run it on your LC/MS system with your tuning parameters under your chromatographic conditions to see if it gives the same spectrum as you obtain in your precursor scan.

APPENDIX D

GLOSSARY OF LC/MS TERMS

Affinity chromatography. Partition chromatography run on a bonded-phase column containing a functional group that attracts and holds a specific compound or class of compounds. Retained compounds are eluted using either pH control or a specific counter compound.

Alpha factor. An HPLC separation factor that is controlled by changes in the chemistry of the mobile phase, the stationary phase in the column, or the injected sample. Temperature also produces an alpha effect in columns not dissolved or degraded by elevated temperature. Also called *chemical factor*.

APCI (atmospheric-pressure chemical ionization). A term used interchangeably with *ion spray ionization*.

API (atmospheric-pressure ionization). Any of the techniques used to evaporate solvent and ionize the compounds injected for mass spectral analysis.

Autosampler. A programmable device that makes possible unattended injection of sample and standards solutions onto the head of the HPLC column.

Base peak. The most intense ion fragment in a compound's spectrum under a given set of experimental conditions.

C18–silica column. The single most commonly used HPLC column. Usually, has an octyldecyl group bound by a silane function to the underlying silica particle. C18 columns are also available with octyldecyl groups bound to zirconium and polymer particles.

Calibration standards. Compounds injected either through the interface or from the LC, to allow calibration or autotuning of a mass spectrometer's mass axis to the *m/z* peak position and heights expected.

LC/MS: A Practical User's Guide, by Marvin C. McMaster
Copyright © 2005 John Wiley & Sons, Inc.

Capillary zone electrophoresis (CZE). A separation technique based on movement of ionized compounds through a capillary tube filled with buffer toward a high voltage of the opposite polarity. Separation is based on the compound's size and charge potential.

Check valve. A ball-and-seat mechanism in the inlet and outlet fittings of an HPLC pump that prevents reverse flow of solvent and allows pressurization of the column to overcome backpressure from packing material flow resistance. Check valve contamination is a major source of pumping problems. *See also* Pacification.

CI. Chemical ionization in GC/MS, in which the MS source is flooded with a low-molecular-weight gas such as carbon dioxide to produce a much softer ionization, generally leading to retention of the molecular ion for molecular-weight determination.

Column. A packed tube filled with a stationary-phase particles used to achieve liquid chromatography separations. Different bonded phase can be attached to the stationary phase to achieve partition, ion exchange, affinity, or size separations.

Column bridge. A device made of two unions connected with compression fittings to the ends of a 5-ft section of 0.01-in.-ID tubing. It is used to replace the column for systems washing and system pacification.

Data/control system. The "brains" of the LC/MS system; programs and starts system components and controls MS scanning and lens, acquiring and processing the data from the detector.

Detector. A device that produces a voltage change in response to a change in the composition of the material in its flow cell.

Differential pumping. An arrangement in which two chambers connected by a small orifice (e.g., an MS source and analyzer) have two pump connections through different-diameter exhaust tubes, capable of providing different pumping rates and vacuums in the two chambers.

Efficiency factor. An HPLC separation factor that measures the sharpness of chromatographic peaks. It is controlled by column length and flow rate and is influenced by particle shape, diameter, and column packing efficiency.

EI. Electron ionization using a 70-eV filament in a mass spectrometer's source. Uncharged materials from the GC are ionized to a molecular ion that usually fragments into daughter ions.

Electrospray. An interface placed between the LC and MS modules to evaporate effluent and ionize eluted compounds. Electrospray passes the column output through a heated, electrically charged tube, usually aided by a nebulizing gas. It is the interface of choice for polar or charged samples such as proteins.

Fourier transform LC/MS (LC/FT-MS). A separation technique in which an LC sample is ionized in an interface, injected into a mass spectrometer, held in place by a cyclonic trapping voltage, excited to a higher orbit by a "chirping" multifrequency signal, and transmits an RF signal characteristic of all the masses present. Computer transformation of the multifrequency signal allows plotting of intensity versus *m/z* spectra with very high sensitivity at each chromatographic point. The sample is not destroyed, so the signal can be sampled repeatedly to produce a very accurate sample measurement.

Gas chromatography. A separation technique in which the volatile analyte is swept by a carrier gas down a column packed with packing coated with an absorbing liquid. Differential partition between the two phases by sample components leads to band separation and elution into a detector. Often combined with a mass spectrometer to create a GC/MS system.

Hybrid MS/MS. A combination of two or more mass spectrometer analyzers of different types. Used like a triple quadrupole system to separate and then fragment compounds in the LC effluent to aid in identifying a compound's structure from its fragmentation ions.

Injector. A device used to move a sample in undiluted form onto the head of the HPLC column. *See also* **Autosampler**.

Internal standard. A compound added to all analyzed samples in equal concentration during the last dilution before sample injection. Its purpose is to correct for variations in sample injection size. It can also be used to correct for variations in peak retention times.

Ion-exchange chromatography. Partition chromatography in which the column has a bonded phase with a permanent or inducible charge opposite to that which attracts the target molecules. Compounds are eluted either with a counterion, salt, or pH gradients.

Ion spray interface. An atmospheric-pressure ionizing interface between the LC and MS modules. Column effluent is nebulized or entrained with an inert gas in a capillary tube and sprayed across a coronal discharge needle operating at high voltage to evaporate solvent and volatile buffer and ionize the sample. It is the interface of choice for neutral and nonpolar samples.

Ion trap detector (ITD). A desktop three-dimensional spherical-segment ion trap MS that ionizes and holds the ionized sample within a circular electromagnet until swept with a dc/RF frequency signal that releases the ionized sample into the ion detector.

Linear ion trap (LIT). A linear ion trap mass spectrometer is based on a quadrupole analyzer with end electrodes for trapping or releasing ions. It has a much higher loading capacity than that of a three-dimensional spherical ion trap, which translates to more accurate measurements for trace analysis.

Mobile phase. The liquid phase being pumped through an HPLC column to act as a carrier for the sample, to establish equilibration with the stationary phase, and eventually to elute the separated compound peaks.

Molecular weight. Summation of the weights of all the elements in a molecule, expressed in amu or dalton. In mass spectral analysis, the weight of the molecular ion produced in a soft chemical ionization such as ion spray or electrospray.

MS/MS system. A tandem, triple quad, or hybrid dual mass spectrometer system used to study MS fragmentation mechanisms. The second analyzer is used to further fragment ions separated in the first analyzer. Separation of the fragments can be done by scanning the trapping analyzer or in a third analyzer.

m/z.   Mass divided by charge measured in amu or dalton. The x-axis for a mass spectrum, indicating that MS spectra are dependent on both the ion's mass and the charge on the ion.

Neutral loss experiment.   A MS/MS experiment in which the first analyzer measures the precursor ion fragments while the second analyzer looks for product ions that have a constant mass difference, the mass of the neutral loss product molecule formed during the fragmentation. Only precursor ions that form this neutral compound will show in the product ion scan that is produced.

Normal-phase chromatography.   Partition separation on unbounded polar silica columns. Mobile phase is usually a nonpolar solvent but can be an acidic polar solvent.

Pacification.   Washing a system with 6 N nitric acid after removing the HPLC column. Cleans the pump heads, check valves, and the injection loop and protects them against salt corrosion for a period of time.

Plunger and seal.   Solvent flow in an HPLC pump is created by a sapphire plunger driven by a reciprocating cam-and-spring mechanism through a Teflon seal into a pumping chamber. Check valves in the pump's inlet and outlet fitting prevent reverse solvent flow and allow pressure buildup to overcome flow resistance in the column.

Precursor ion.   The target ion submitted for fragmentation studies in a MS/MS system. A precursor scan is a mass scan of the fragmentation ions produced by thermal collision with a heavy gas molecule.

Quadrupole analyzer.   Mass spectrometer analyzer based on four circular rods held in a hyperbolic configuration and swept with a variable-frequency dc/RF signal, allowing selection of individual mass fragments.

Resolution equation.   A measure of a column's separating power. It combines retention, separation, and efficiency factors into a single equation and shows their interactions.

Retention factor.   An HPLC column resolution factor measuring how separation is affected by residence time on the column. Retention is controlled by mobile-phase polarity and to a certain degree by column temperature.

Retention time.   The length of time a compound stays on the column under a given set of experimental conditions.

Reverse-phase chromatography.   Partition chromatography in which the column has a nonpolar bonded phase and the mobile phase uses polar solvents mixed with water. The majority of reverse-phase separations are made on C18 bonded phase on silica columns.

Ring electrode.   The central electrode of a three-dimensional ion trap, used to hold ion fragments in circular orbits until time to elute them into the detector.

Roughing pump.   The first pump in a vacuum system, used to reduce pressure initially from atmospheric pressure to a low pressure that can serve as a starting point for the high-performance vacuum pump. Today's roughing pump is usually a mechanical rotary vane oil pump.

Scan. A mass spectrometer operational mode in which the amount of each mass unit is measured by continuously changing the dc/RF frequency on a quadrupole. Mass can be scanned low to high or high to low, the latter leading to less intermass tailing and more accurate relative height measurements.

Secondary detector. A detector, usually a UV detector, connected to an effluent diverter in the ionizing interface to take the excess effluent from the MS measurement and provide a second measurement of the chromatography peaks by an alternative form of detection.

Separation factor. A column resolution factor controlled by the column's chemistry and by temperature. Changes in this factor result in shifting of relative peak positions. *See also* **Alpha factor** and **Resolution equation**.

SIC (single-ion chromatogram). A chromatogram produced by displaying the ion current produced versus time for a given mass (m/z). It can be produced by operating in single-ion mode or extracted from a scanned fragment database.

SIM (single-ion monitoring). The mass spectrometer measures one or only a few specific m/z points. Since fewer mass measurements are made than in scan, they are made more often, with a proportional increase in sensitivity. Also an acronym for selective ion monitoring.

Size-separation chromatography. Chromatography in which separation is made on the size of the molecules on controlled-pore-size columns. Also referred to as *gel filtration*.

Spectra. Bar plots of signal intensity in volts versus ion fragment mass/charge, measured in amu, for a given MS scan or range of scans. The data are usually summed around unit mass and presented as a bar graph of intensities relative to the base peak.

SPE. Sample filtration and extraction cartridge columns used for preinjection filtration and chromatographic cleanup of a sample. SPE columns are available for all separation modes used in LC/MS.

Supercritical fluid chromatography (SCF). A column separation technique using pressure and temperature control to convert a gas into a fluid that is used as the mobile phase for liquid/solid chromatography. Sample recovery is made by releasing the pressure to turn the mobile phase back into a gas.

Surrogate. A standard compound added in known amounts to all processed samples. Its purpose is to detect and correct for sample loss due to extractions and handling errors. Usually, it is a deuterated or other isotopic-labeled derivative of an analyzed compound not normally found in nature.

Target compound quantitation. Quantitation based on identifying a compound from its precursor scan by locating target and qualifier ion fragments. Once identified, the target ion signal strength is compared to known amounts of standards to determine the amount present.

Target ion. One of a compound's MS ion fragments chosen for identifying and quantitating the amounts of the compound present in mixtures of standards and unknowns.

Temperature ramp.   A gradual controlled increase of temperature with time. It is used in combination with holds and other ramps in building a column heater temperature program for resolving compounds on a zirconium column.

TIC (total ion chromatogram).   A chromatogram produced by measuring the total ion current from a mass spectrometer versus time. A TIC data point represents a summation of all mass fragments present at a given time.

Time-of-flight LC/MS (LC TOF/MS).   A chromatographic technique in which an MS detector analyzes effluent by ionizing it with pulse laser energy bursts and identifying mass fragments by using the time they take to travel a flight tube and reach a detector. LC/TOFMS is becoming popular in the analysis of charged biochemicals, proteins, and DNA restriction fragments with multiple charges.

Triple-quad LC/MS/MS.   A tandem quadrupole system in which an LC feeds a mass detector with three quadrupole units in series. The second quadrupole acts as a holding and collision cell in which fragments separated in Q1 can interact with a heavy gas such as xenon and fragment for separation in Q3. Used primarily for studying fragmentation mechanisms to aid in precursor compound structure identification.

Turbomechanical pump.   A high-vacuum pump that uses a series of vanes, mounted on a shaft, that rotate rapidly between stator plate entraining air molecules, dragging them out of the volume evacuated. A turbo pump operates like a jet engine to evacuate a mass spectrometer to the high vacuum (10^{-5} to 10^{-7} torr) needed for operation.

Void.   A channel inside the packing of an HPLC column that ruins the separation. Voids can open spontaneously when a column is shocked or flow is reversed. They can also be caused by fines breaking off irregular packing or by silica packing dissolving at pH above 8.0.

Void volume.   A measure of the total solvent volume inside a particle-packed HPLC column. This value is used in the resolution equation to normalize retention values to make them independent of column length. Also used to define washout volume for column regeneration.

APPENDIX E

LC/MS SELECTIVE READING LIST

JOURNALS

American Journal of Mass Spectrometry
American Laboratory
Analyst
Analytical Biochemistry
Analytical Chemistry
Environmental Science and Technology
Journal of Pharmaceutical and Biomedical Analysis
LC/GC
Rapid Communications in Mass Spectrometry

BOOKS

Ardrey, Bob, *Liquid Chromatography–Mass Spectrometry: An Introduction*, Wiley, Chichester, West Sussex, England, 2003, 276 pp.

Barcello, Damia, ed., *Applications of LC-MS in Environmental Chemistry*, Elsevier Science, New York, 1996.

Lee, Mike S., *LC/MS Applications in Drug Development*, Wiley-Interscience, New York, 2002, 243 pp.

McLafferty, Fred W., and Turecek Frantisek, *Interpretation of Mass Spectra*, 4th ed., University Science Books, Mill Valley, CA, 1993, 377 pp.

McMaster, M. C., *HPLC: A Practical User's Guide*, VCH Publishers, New York, 1994, 211 pp.

McMaster, Marvin, and Christopher McMaster, *GC/MS: A Practical User's Guide*, Wiley-VCH, New York, 1998, 167 pp.

Snyder, L. R., and J. J. Kirkland, *Introduction to Modern Liquid Chromatography*, Wiley, New York, 1979, 863 pp.

Synder, L. L., et al., *Practical HPLC Methods Development*, Wiley, New York, 1997, 299 pp.

Watson, J. Throck, *Introduction to Mass Spectrometry*, 2nd ed., Raven Press, New York, 1985, 351 pp.

Willoughby, Ross, et al., *A Global View of LC/MS*, Global View Publishing, Pittsburg, 2002.

PAPERS

Allen, Mark, and Bob Shushan, "Atmospheric Pressure Ionization–Mass Spectrometry Detection for Liquid Chromatography and Capillary Electrophoresis," *LC/GC*, 11(2), 112–126, 1993.

Amirav, Aviv and Ori Granot, "LC-MS with Supersonic Molecular Beams," *J. Am. Soc. Mass. Spectrom.*, 11, 587–591, 2000; www.tau.ac.il/chemistry/amirav/lcms.shtml.

Chalk, Rod, and Urs Matter, "Analysis of High-Mass Biopolymers: Applications and Advantages of Cryodetector Mass Spectrometry," *Am. Biotech. Lab.*, April 2004, 26–28.

Duncan, W.P., and Perkins, P.D. "LC-MS With Simultaneous Electrospray and Atmospheric Pressure Chemical Ionization," *Am. Lab.*, March 2005, 28–33.

Henry, Richard A., "Highly Selective Zirconia-Based Phases for HPLC Applications," *Am. Lab.*, November 2002, 18–25.

Henion, Jack, et al., "High-flow Ion Spray Liquid Chromatography/Mass Spectrometry." *Anal. Chem.*, 65, 439–446, 1993.

Hopfgartner, G., et al., "High-Flow Ion Spray Liquid Chromatography/Mass Spectrometry," *Anal. Chem.*, 65, 439–446, 1993.

Sandra, Pat, et al., "Consideration on Column Selection and Operating Conditions for LC-MS, Interfaces for LC-MS, Mass Analysers for LC-MS," *LC/GC Europe*, December 2001; www.lcgceurope.com.

Smith, Richard D., "Trends in Mass Spectrometry Instrumentation for Proteomics," *Trends Biochem.*, 20(12), 2002, S3–S7; www.trends.com.

Tiller, P. R., et al., "Qualitative Assessment of Leachables Using Data-Dependent Liquid Chromatography/Mass Spectrometry and Liquid Chromatography/Tandem Mass Spectrometry," *Rapid Commun. Mass Spectrom.*, 11, 1570–1574, 1997.

Zahleen, K., et al., "Screening Drugs of Abuse by LC/MS," Agilent Technologies, publication 5889-1267EN, 2004, 4 pp.; www.agilent.com/chem.

INDEX

Printed and bound by CPI Group (UK) Ltd, Croydon, CR0 4YY

16/04/2025

14658417-0005